Pottery and Ceramics

ART AND ARCHITECTURE INFORMATION GUIDE SERIES

Series Editor: Sydney Starr Keaveney, Associate Professor, Pratt Institute Library, Brooklyn, New York

Also in this series:

AMERICAN ARCHITECTS TO THE FIRST WORLD WAR—*Edited by Lawrence Wodehouse*

AMERICAN ARCHITECTS FROM THE FIRST WORLD WAR TO THE PRESENT—*Edited by Lawrence Wodehouse*

AMERICAN DECORATIVE ARTS AND OLD WORLD INFLUENCES—*Edited by David M. Sokol**

AMERICAN DRAWING—*Edited by Lamia Doumato**

AMERICAN PAINTING—*Edited by Sydney Starr Keaveney*

AMERICAN SCULPTURE—*Edited by Janis Ekdahl*

ANCIENT EGYPTIAN ART AND ARCHITECTURE—*Edited by Eleanore Wedge**

ART EDUCATION—*Edited by Clarence Bunch*

BRITISH ARCHITECTS, 1840-1976—*Edited by Lawrence Wodehouse*

COLOR THEORY—*Edited by Mary Buckley*

HISTORIC PRESERVATION—*Edited by Arnold L. Markowitz**

INDUSTRIAL DESIGN—*Edited by J. Roger Guilfoyle**

INDIGENEOUS ARCHITECTURE WORLDWIDE—*Edited by Lawrence Wodehouse**

EUROPEAN PAINTING, 1870-1910—*Edited by Timothy Daum**

STAINED GLASS I—*Edited by William Serban and Darlene A. Brady**

STAINED GLASS II—*Edited by William Serban and Darlene A. Brady**

TWENTIETH-CENTURY EUROPEAN PAINTING—*Edited by Ann Marie Bergholtz**

VICTORIAN ART: ARCHITECTURE, SCULPTURE, DECORATIVE ARTS—*Edited by Marlene A. Palmer**

VICTORIAN ART: PAINTING, DRAWING, GRAPHIC ART—*Edited by Marlene A. Palmer**

*in preparation

The above series is part of the

GALE INFORMATION GUIDE LIBRARY

The Library consists of a number of separate series of guides covering major areas in the social sciences, humanities, and current affairs.

General Editor: Paul Wasserman, Professor and former Dean, School of Library and Information Services, University of Maryland

Managing Editor: Denise Allard Adzigian, Gale Research Company

Pottery and Ceramics

A GUIDE TO INFORMATION SOURCES

Volume 7 in the Art and Architecture Information Guide Series

James Edward Campbell

Ceramics Instructor
Friends Academy
Locust Valley, New York

Gale Research Company
Book Tower, Detroit, Michigan 48226

Library of Congress Cataloging in Publication Data

Campbell, James Edward.
Pottery and ceramics.

(Art and architecture information guide series; v.7)
Includes indexes.
1. Pottery—Bibliography. I. Title.
Z7179.C35 [NK4225] 016.738 74-11545
ISBN 0-8103-1274-3

To My Parents

James and Loretta Campbell

VITA

James Edward Campbell has been an instructor in ceramics at Friends Academy in Locust Valley, Long Island, since 1972. He is a graduate of the Catholic University of America, where he received a B.A. in English literature, and of Pratt Institute, where he received an M.F.A. in ceramics. It was while working as graduate assistant in ceramics at Pratt Institute that he began research for this project. He is a member of the American Crafts Council and of the Long Island Craftsman's Guild.

CONTENTS

Contents

PREFACE

This bibliography is intended for the use of researchers, students, and individuals who have some basic knowledge of ceramic history and/or some technical knowledge of ceramic materials. For the most part, these references are fairly specialized. While answering the reference needs of a very broad range of readers, my text should be of principal benefit to the ceramic historian, collector, and artist. Various chapters cite archaeological and ethnological materials; other chapters guide the reader to industrial and architectural information. Although the latter features may be only of peripheral interest, I have incorporated them in the hope that they may be of value to some researchers and serve as a source for further information.

The only published bibliography on ceramics of equivalent scope appears to have been M.L. Solon's CERAMIC LITERATURE: AN ANALYTICAL INDEX (see chapter 1), described by its author as including all works published in all languages in the ceramic arts. Since that book was published in 1910, an updated work in the field now seems long overdue.

Even allowing Solon a measure of literary license in his claim of completeness, I make no such claim concerning the overall scope of my book. For the most part only works written in English have been included. The quantity of ceramic material available today makes a study of Solon's scope staggering, if not clearly impossible. In the area of English pottery and porcelain alone, the number of popular monographs published over the years would probably number in the hundreds. As a result, I have omitted many early survey-type volumes if their subjects appear to me to have been better covered by more recent treatments or if they appear to have no really distinguishing characteristics. With regard to the latter condition, however, I have taken a somewhat conservative approach so that, if the text in question seemed of even limited value, I included it.

Virtually all of my citations deal exclusively with ceramics. Occasionally, I have cited non-English-language publications if they contain exceptional illustrations or provide an unusual approach to their subjects. Occasionally, too, I have resorted to including less adequate references if they are the only available sources of information on a particular ceramic ware.

In the October 1975 issue of CRAFT HORIZONS Florence Pettit discusses "What's Wrong with the Craft Books Boom!" She criticizes the glut of materials on crafts, including ceramics, flooding the market today. Much that is written is of poor quality and repetitious. While I have admitted a broad range of ceramic material into chapter 11, my natural prejudice has tended more toward serving the professional craftsperson than the hobbyist, and I therefore have made no attempt to include every one of the latest ceramic crafts books on the market.

While the title POTTERY AND CERAMICS may appear redundant to many readers, I have used it on good advice simply to avoid any ambiguity concerning the scope of my text. The word *pottery* has traditionally been used to define the low-fired and often unglazed earthenwares while the word *ceramics* has usually characterized the high-fired and glazed wares such as stoneware and porcelain. Since the pottery wares of ancient and nontechnical societies have been included in this publication, I have used the dual title; the word *ceramics*, however, is commonly used today as encompassing all of the clay arts. For the most part, it is in this general and popular sense that the term *ceramics* has been broadly applied throughout this text.

Many of the chapters begin with very brief introductions of one to several sentences in length. The reader is advised to glance at these introductory remarks since they were designed to summarize quickly the scope, content, and organization of that chapter *in situ*, and such information would almost certainly go unnoticed in a longer and more detailed preface.

Additional aids have been incorporated for the benefit of the nonspecialist using this guide. The numbers from Mary Chamberlin's GUIDE TO ART REFERENCE BOOKS (see entry under "Reference Works") have been included in my annotations of appropriate entries throughout my text. In situations where a reference deals with several overlapping general areas of ceramic art, or overlapping periods, I have sometimes cross-referenced the publication. This technique has been minimized, however, so that such referrals do not become more of an annoyance than a help to the reader. Asterisks have been placed beside a small selection of works that would be suitable for a small, nonspecialized library. The books so marked include a number of works published after Chamberlin's bibliography and represent a broad, general survey of the history and craft of ceramics. Specialized studies and the most simplified popularizations have been excluded from this selection.

One of the difficulties of preparing a critical bibliography is that the variety of the material covered does not allow the writer to feel completely confident of his evaluations. Unless he is closely connected to a particular segment of a field of knowledge, it is sometimes perplexing to determine whether an old volume is a classic text that has historical as well as present-day usefulness or whether it is simply a massive old tome that is hopelessly dated. I do feel that my judgments throughout this volume are reasonable and valid, but I am not so naive as to deny the possibility of occasional glaring inaccuracies. My only defense is that whenever corrective information later came my way I made appropriate changes in the manuscript. There literally might be no end to these revisions, however, and since I had neither the desire nor the resources to turn this project into a lifetime of research, I completed it, in the hope that its merits would outweigh its deficiencies.

My research for this project was done primarily at the libraries of Pratt Institute, Columbia University, and the American Crafts Council. Additional work was completed at the World Crafts Council, the Metropolitan Museum of Art, the American Museum of Natural History, and with Sarah Bodine at Watson-Guptill Publications. I would like to express my sincerest thanks to these institutions for the use of their research facilities and to the individuals there for their generous help in finding material. Virtually without exception busy people gave generously of their time, put other work aside, and made every effort to solve any difficulties that arose.

Many thanks are also due to the great number of people who gave of their time and knowledge in offering advice on this guide. I have no doubt that in some ways many of my most accurate judgments are the result of their greater expertise.

The help of three individuals really has made this work possible. Sydney Keaveney, the editor of this series, has consistently been a source of encouragement in every phase of this project's production. Cathy Bouchard typed the manuscript, and I can truthfully say that without her help this guide would never have been completed. Finally, Peggy Brucia volunteered for the long and tedious job of proofreading what must be one of the most difficult types of manuscripts with which to deal. To all three I am most grateful.

Chapter 1
REFERENCE WORKS

With the exception of Solon's INDEX, none of the texts included in this chapter deals exclusively with ceramics. Rather, each of these reference works contains some ceramic information which may be useful to the reader for further research and study.

*AMERICAN ART DIRECTORY. Edited by Jaques Cattell Press. New York and London: R.R. Bowker Co., 1952-- . Irregular.

Published most recently on a biennial basis, this directory lists American art organizations and musuems and briefly describes their activities and/or collections. It also lists art schools in the United States, Canada, and abroad, art magazines, and other art-related material. This serial is a continuation of the AMERICAN ART ANNUAL, published from 1898 to 1948.

ARTBIBLIOGRAPHIES: MODERN. Santa Barbara, Calif. and Oxford, Eng.: American Bibliographical Center--Clio Press, 1973-- . Semiannual.

This computerized bibliography contains annotated references to books, periodical material, and exhibition catalogs encompassing material from 1800 to the present. There is a separate ceramic art section and a substantial number of cross-referenced sources. It is a continuation of LOMA (LITERATURE OF MODERN ART), published in 1969, 1970, and 1971.

ART INDEX. New York: H.W. Wilson Co., 1929-- . Quarterly, annual cumulations since 1967 (previously biennial and triennial).

Covering approximately two hundred periodicals, this index is "a cumulative author and subject index to fine arts periodicals and museum bulletins covering archaeology, architecture, arts and crafts, ceramics, decoration and ornament, graphic arts, industrial design, interior decoration, landscape architecture, painting and sculpture." Technical material as well as information on historical and contemporary ceramic appreciation are included in the pottery and ceramics sections. There are also a number of cross-references.

Reference Works

BOOK REVIEW INDEX. Detroit: Gale Research Co., 1965-- . Bimonthly (cumulated annually).

This index identifies new books reviewed in over two hundred periodicals. The alphabetical listing of entries by author and the lack of subject index somewhat limit the usefulness of this text.

*Chamberlin, Mary W. GUIDE TO ART REFERENCE BOOKS. Chicago: American Library Association, 1959. xiv, 418 p. Appendix, pp. 345-60. Index, pp. 361-418.

This volume lacks a number of valuable new texts in ceramics. However, Chamberlin's book remains one of the most commonly consulted art bibliographies. Because it must deal with such a broad range of material, only a small part could be devoted to ceramics; some fifty-two volumes are listed and annotated. The code number of each of the ceramic books listed by Chamberlin has been included in the annotations of this bibliography.

Chicago. The Art Institute of. INDEX TO ART PERIODICALS, COMPILED IN THE RYERSON LIBRARY OF THE ART INSTITUTE OF CHICAGO. 11 vols. Boston: G.K. Hall, 1962.

Admitting material back to 1907, this index incorporates both periodicals and museum publications. Some of these were later taken over by the ART INDEX while others are not included. Although a number of pottery references are contained here, the majority of these articles date from the 1920s and 1930s.

_____. INDEX TO ART PERIODICALS, COMPILED IN THE RYERSON LIBRARY OF THE ART INSTITUTE OF CHICAGO. Supp. 1. Boston: G.K. Hall & Co., 1975. 573 p.

While this supplement does contain a number of pottery references, many of the entries are taken from material in non-English periodicals. The value of this work for ceramic information is very much secondary to that of the ART INDEX.

Clapp, Jane. ART IN LIFE. New York: Scarecrow Press, 1959. 504 p.

This index includes a "selective listing of photographs of architecture, sculpture, and decorative arts, such as ceramics" published in LIFE MAGAZINE from 1936 through 1956. Some fifty ceramic references touch upon a wide range of ceramic history. It should be noted that the examples used have been selected principally for their illustrations.

_____. MUSEUM PUBLICATIONS. Part 1: Anthropology, Archaeology and Art. New York: Scarecrow Press, 1962. 434 p. Index, pp. 294-434.

This bibliography does not include periodicals or regularly printed museum publications, such as bulletins, etc. Rather, its emphasis is on independently printed books, pamphlets, and monographs available from 276 American and Canadian museums. There is a small amount of ceramic material related to both archaeology and the decorative arts, and the text does offer information on museum guides, catalogs, etc.

*CONTEMPORARY CRAFTS MARKET PLACE. Compiled by the American Crafts Council. 1975-1976 Edition. New York and London: R.R. Bowker Co. Irregular. Index. Paper.

This guide lists and describes briefly shops, galleries, organizations, courses, audiovisual materials, suppliers, shippers, insurers, and publications involved in various crafts, including ceramics. It encompasses an enormous amount of craft material and is a valuable source book for anyone interested in contemporary American ceramics.

Davis, Alexander. LOMA (LITERATURE OF MODERN ART). 3 vols. London: Lund Humphries, 1969, 1970, 1971.

See: ARTBIBLIOGRAPHIES: MODERN (above, this section).

INTERNATIONAL DIRECTORY OF THE ARTS. 2 vols. Frankfurt, W. Ger.: Verlag Muller Gmbh. & Co., 1949-- . Biennial. Index.

While this directory does not have a separate ceramic section, the comprehensive nature of its listings of museums, galleries, dealers, art experts, restorers, and auctioneers makes it a valuable source of information. The listings of a number of restorers, for example, specify the medium in which they specialize, such as porcelain.

IRREGULAR SERIALS AND ANNUALS: AN INTERNATIONAL DIRECTORY. New York and London: R.R. Bowker Co., 1967-- . Biennial. Index.

The 1974-75 edition of this directory identifies and describes approximately forty-six English- and non-English-language serial publications under the subject key of "Ceramics, Glass and Pottery."

THE LIBRARY JOURNAL BOOK REVIEW. New York and London: R.R. Bowker Co., 1968-- . Annual.

Covering a small number of ceramic volumes under "Decorative Arts and Crafts," the JOURNAL BOOK REVIEW evaluates new, general interest books on the history and craft of ceramics.

*Lucas, E. Louise. ART BOOKS: A BASIC BIBLIOGRAPHY ON THE FINE ARTS. Greenwich, Conn.: New York Graphic Society, 1968. 245 p. Index, pp. 207-45.

This volume combines and revises two earlier works by Lucas, BOOKS ON ART: A FOUNDATION LIST, first published in 1929; and THE HARVARD LIST OF BOOKS ON ART, published in 1952. These publications were intended as a basic selection of works suitable for an undergraduate college library. While ceramics is treated only briefly, its thirty references are basic texts encompassing almost every facet of ceramic history.

MASTER'S THESES: CRAFTS. New York: American Crafts Council, 1976. 153 p. List of Schools in the Directory, pp. 149-51. Other Schools Related to the Directory, pp. 151-53. Paper.

This directory includes fifty pages of theses topics in the field of ceramics. Studies of ceramic materials, equipment, and techniques as well as studies of form, design, and ceramic education are represented. Each entry identifies the title of the thesis, its author, date, present location, and the degree for which it was submitted.

THE OFFICIAL MUSEUM DIRECTORY. Washington, D.C.: American Association of Museums; Skokie, Ill.: National Register Publishing Co., 1961-- . Biennial.

This directory briefly identifies the personnel, collections, activities, hours, and admission charges of museums in the United States and Canada. Some 5,000 institutions are included.

*READERS' GUIDE TO PERIODICAL LITERATURE: AN AUTHOR AND SUBJECT INDEX. New York: H.W. Wilson Co., 1900-- . Semimonthly (cumulated quarterly and annually).

This guide is "a cumulative author subject index to periodicals of general interest published in the United States." Ceramic material is included under the headings of ceramic sculpture, glazes, and pottery. CERAMICS MONTHLY, CRAFT HORIZONS, and ANTIQUES are among the periodicals indexed. There is occasionally some overlapping of material here with that in the ART INDEX.

Solon, M.L. [Louis Marc]. CEREMIC LITERATURE: AN ANALYTICAL INDEX. London: Charles Griffin & Co., 1910. 660 p. (Chamberlin, no. 1735).

According to the author, this index includes all the works published in all languages on the history and technology of ceramic art. Part 1 of this volume is an alphabetical listing of all these works by author, while part 2 repeats in an abridged form the same works classified by subject, country of origin, and date of publication. The second part also includes sections on ceramic technology, tiles, architectural terra-cotta, museums, collections, and exhibitions. While Solon's work is quite extensive, "all works in all languages" should be understood to mean works in

European languages. Although this study's coverage is very extensive, its early date reduces its usefulness to readers with somewhat specialized interest.

THE STANDARD PERIODICAL DIRECTORY. Edited by Leon Garry. New York: Oxbridge Publishing Co., 1963-- . Biennial.

Collected under the section entitled "Glass, Stone and Clay," the 1973 edition of this directory describes (briefly) approximately thirty ceramic publications. Most of these publications are of an industrial or scientific nature.

ULRICH'S INTERNATIONAL PERIODICALS DIRECTORY. New York and London: R.R. Bowker Co., 1932-- . Biennial. Index.

This directory identifies approximately one hundred current English- and foreign-language periodicals on ceramics and pottery.

THE WORLDWIDE ART CATALOGUE BULLETIN. Boston: Worldwide Books, 1962-- . Quarterly.

This sales bulletin offers a selection of the more important art catalogs recently issued as well as older ones previously unavailable. Catalogs of both historical and contemporary ceramics are identified, each with a brief description, and indexed by title, artist, chronology, mediums, and subject.

"World Crafts Directory." 4 vols. New York: World Crafts Council, 1973 (Asia, Africa, Latin and North America), 1975 (Europe). Europe, 94 p. Asia, 64 p. Africa, 54 p. Latin and North America, 53 p. Irregularly revised (semiannually). Paper. Mimeographed.

Originally published in 1964 as A SHORT GUIDE TO WORLD CRAFTS, this directory contains the names and addresses of craft organizations, associations, museums, schools, shops, and craft publications in some ninety countries throughout the world. These volumes do not indicate whether the publications are in English or have English summaries, and some of the material is listed without any annotation. On the whole, however, the coverage of world crafts here is extensive and quite valuable.

Chapter 2
GENERAL HISTORIES, DICTIONARIES, AND ENCYCLOPEDIAS

Barber, Edwin Atlee. THE CERAMIC COLLECTOR'S GLOSSARY. New York: printed for the Walpole Society, 1914. 119 p. Figs. (Chamberlin, no. 1736).

This short glossary identifies a variety of shapes, ornamental styles, and surface decorations by name to establish a "uniform ceramic nomenclature." Some types of wares such as Wedgwood and Bennington are also identified.

_____. LEAD GLAZED POTTERY. Part 1: (COMMON CLAYS) PLAIN GLAZED, SGRAFFITO AND SLIP-DECORATED WARES. Art Primer. Pennsylvania Museum and School of Industrial Art. New York: Doubleday, Page & Co., 1907. 34 p. 47 illus.

Part of a series published for the Pennsylvania Museum, this text briefly examines the characteristics, origins, and decoration of earthenware pottery.

*Berendsen, Anne, et al. TILES: A GENERAL HISTORY. Translated by Janet Seligman. New York: Viking Press, 1967. 286 p. Illus. Bibliography, pp. 273-79. Index, pp. 281-86.

This well-illustrated popular history covers the whole range of tiles from those of antiquity to more recent times. Also included is a brief account of the materials and methods of tile production.

*Boger, Louise Ade. THE DICTIONARY OF WORLD POTTERY AND PORCELAIN. New York: Charles Scribner's Sons, 1971. 533 p. 60 color pls., 550 halftones, 150 figs., 550 marks. Bibliography, pp. 525-33.

This handsome volume contains material from prehistoric times to the present and includes some 2,200 entries on processes, factories, styles, terms, and individuals. There are notes on each of the many illustrations and entries are cross-referenced.

Brooklyn Institute of Arts and Sciences. The Brooklyn Museum. THE ART & TECHNIQUE OF CERAMICS: AN EXHIBITION PRESENTED BY THE ROCKEFELLER FOUNDATION INTERNES OF THE BROOKLYN MUSEUM. Brooklyn,

N.Y.: 1937. 112 p. Illus. Bibliography, pp. 109-10.

The stated purpose of this work is to present a survey of the three major stages in the advancement of ceramic art, namely porous pottery, glazed porous pottery, and vitreous ware. The very brief text is organized geographically, and the major portion of the volume is composed of halftone illustrations.

Burgess, Fred W. OLD POTTERY & PORCELAIN. The Home Connoisseur Series. London: George Routledge & Sons, 1916. xvii, 426 p. 130 halftones. Glossary, pp. 415-17. Index, pp. 419-26.

Cursorily treating the origins, techniques, and early styles of pottery, Burgess devotes most of this volume to the factories and styles of English and other European ceramics. Oriental pottery and porcelain, tiles, terra-cottas, and historic pottery are also briefly treated.

*Burton, William. A GENERAL HISTORY OF PORCELAIN. 2 vols. London: Cassell & Co., 1921. Vol. 1: xviii, 204 p. Vol. 2: ix, 228 p. Bibliography, pp. 207-8. Index, pp. 209-28. 32 color pls., 80 halftones. (Chamberlin, no. 1741).

This fairly extensive popular history includes the porcelain of China, Japan, and Persia as well as that of France, Germany, and England. As its title implies, this volume is more concerned with a general description of the different styles of porcelain than with a detailed treatment of names, dates, discoveries, and comparisons.

Chaffers, William. THE KERAMIC GALLERY. 2d ed., revised and edited by H.M. Cundall. London: Gibbings & Co., 1907. xxii, 468 p. 565 illus. (5 color). Index, pp. 467-68. (Chamberlin, no. 1745).

This revised edition of Chaffers's original two-volume publication of 1872 provides a pictorial companion to his MARKS & MONOGRAMS ON POTTERY AND PORCELAIN. Included here are descriptions and examples of ancient pottery, majolica, continental faience and porcelain, English pottery and porcelain, and the wares of both the Near and Far East. For many years Chaffers's texts have been looked upon as the standard reference texts for ceramic collectors.

Charleston, Robert J., ed. WORLD CERAMICS. New York: McGraw-Hill Book Co., 1968. 352 p. Illus. (incl. 64 color pls.). Glossary, p. 345. Glossary of marks, pp. 346-47. Index, pp. 348-52.

Profusely illustrated, this oversize volume covers both Eastern and Western ceramic history. The contents strongly emphasize the European tradition, although they also briefly treat contemporary work and that of the primitive world. The detailed text contains contributions from some twenty authorities.

*Cooper, Emmanuel. A HISTORY OF POTTERY. New York: St. Martin's Press, 1972. 276 p. Illus. (incl. 8 color pls.). Glossary, pp. 259-64. Bibliography, pp. 265-70. Index, pp. 271-76.

Cooper's well-illustrated popular history briefly treats the development of pottery in both Eastern and Western civilizations from its beginnings to the twentieth century. Directions in contemporary ceramics are touched upon briefly.

*Cottier-Angeli, Fiorella. CERAMICS. Translated by Julian Snelling and Claude Namy. Craft and Art. New York: Van Nostrand Reinhold Co., 1974. 143 p. Illus. (part color). Glossary, pp. 136-37. Bibliography, p. 138.

This introductory history begins with a survey of the ceramic process, briefly explaining clay preparation, forming, various decorative techniques, and kiln firing. The history of ceramics is divided into eight chapters and includes Islamic, pre-Columbian, and contemporary ceramics, as well as the more commonly treated classical, Renaissance, European, and Far Eastern traditions.

*Cox, Warren E. THE BOOK OF POTTERY AND PORCELAIN. Rev. ed. 2 vols. New York: Crown Publishers, 1970. 1,158 p. 3,000 illus., 8 color pls. Marks, pp. 1079-150. Index, pp. 1151-58. (Chamberlin, no. 1746).

Quite detailed and containing numerous small illustrations, this volume is an extensive reference work. Although the author terms it a history, Cox's descriptions and illustrations are particularly suitable for the collector and dealer as an aid to identification.

Cushion, John P. ANIMALS IN POTTERY AND PORCELAIN. London: Cory, Adams & Mackay, 1966. 72 p. 53 illus. in 36 pls. (4 color). Bibliography, p. 67. Index, pp. 69-72.

Intended as no more than an introduction for collectors, this volume briefly deals with the materials, techniques, factories, and modelers involved in the production of ceramic animals and birds in both China and Europe. Although the early work of oriental craftsmen is touched upon, the major part of this text is devoted to eighteenth- and nineteenth-century European products.

Dillon, Edward. PORCELAIN. The Connoisseur's Library. New York: G.P. Putnam's Sons; London: Methuen & Co., 1904. xxxv, 420 p. 49 pls. Bibliography, pp. xxvi-xxxii. Marks, pp. 395-404. Index, pp. 405-20. (Chamberlin, no. 1749).

This is a standard popular history covering both Eastern and Western porcelain. Emphasis is on the more highly decorated ware that was of such popular interest to turn-of-the-century collectors.

Dodd, A.E. DICTIONARY OF CERAMICS.

See: Chapter 11, "Technical Works on Ceramic Materials and Processes" (p. 156).

Eberlein, Harold Donaldson, and Ramsdell, Roger Wearne. THE PRACTICAL BOOK OF CHINAWARE. Rev. ed. Philadelphia: J.B. Lippincott Co., 1948. 440 p. 13 color pls., 189 halftones, drawings. Bibliography, pp. 307-10. Index, pp. 311-20. (Chamberlin, no. 1750).

This book is designed as an introductory guide with essential facts for the amateur collector of chinaware. While examples of both American and Eastern chinaware are included, the emphasis is on the many types of chinaware produced throughout Europe.

Furnival, William James. LEADLESS DECORATIVE TILES, FAIENCE, & MOSAIC . . . IN THE HISTORY, MATERIALS, MANUFACTURE & USE OF ORNAMENTAL FLOORING TILES, CERAMIC MOSAIC AND DECORATIVE TILES & FAIENCE. Staffordshire, Eng.: W.J. Furnival, 1904. xxiii, 852 p. 377 illus. 3 appendices, pp. 819-40. Index, pp. 841-52.

As its rather ponderous title suggests, this huge volume covers a very broad spectrum of infomation on the subject of ceramic tile. The author clearly states, however, that his basic purpose is simply to make generally available a series of leadless glazes to assist in eliminating lead poisoning from the ceramic tile industry.

Hamer, Frank. THE POTTER'S DICTIONARY OF MATERIALS AND TECHNIQUES.

See: Chapter 11: "Technical Works on Ceramic Materials and Processes" (p. 158).

Hannover, Emil. POTTERY & PORCELAIN: A HANDBOOK FOR COLLECTORS. 3 vols. Edited by Bernard Rackham. London: Ernest Benn, 1925. Vol. 1: 589 p. 684 illus., 4 maps. Bibliography, pp. 564-73. 3 indexes, pp. 574-89. Vol. 2: 287 p. 402 illus., 2 maps. Bibliography, pp. 269-72. Indexes, pp. 273-87. Vol. 3: 571 p. 858 illus., 3 maps. Bibliography, pp. 546-54. Indexes, pp. 555-71. (Chamberlin, no. 1752).

This is an extensive and detailed survey of the ceramic arts containing hundreds of small illustrations to aid the collector in identification. While there can be little doubt that some of this material is now dated, these volumes do cover an enormous range of information, including references to Far Eastern ceramics (vol. 2), European and Near Eastern earthenware and stoneware (vol. 1), and European porcelain (vol. 3).

Honey, W[illiam].B[owyer]. THE ART OF THE POTTER: A BOOK FOR THE COLLECTOR & CONNOISSEUR. New York: Beechhurst Press, 1955. xvi, 111 p. 160 halftone pls. Index, pp. 107-11.

Intended more as an anthology of beautiful pots than as a serious historical study, this volume contains in its first half an essay and illustrations on the techniques and aesthetics of pottery and in its second half examines three types of excellence in pottery: Korean ware of the Yi dynasty, early European porcelain, and modern English pottery.

Jervis, W[illiam].P[ercival]. ROUGH NOTES ON POTTERY. Newark, N.J.: W.P. Jervis, 1896. 112 p. Figs., marks. Bibliography, p. 102. Paper.

Written and published by Jervis, this pamphlet touches on a wide variety of pottery subjects. Most of the chapters deal briefly with the work of the various English pottery factories and with several types of European wares while additional essays on the antiquity of pottery, oriental wares, the ceramic process, the Rookwood pottery, and several pages of ceramic marks are also included.

______, comp. THE ENCYCLOPEDIA OF CERAMICS. New York: [W.P. Jervis], 1902. 673 p. Illus. Index, pp. 657-73. (Chamberlin, no. 1738).

Arranged alphabetically, this old text identifies thousands of places, companies, and individuals associated with pottery. The emphasis is on European pottery and information on marks and monograms is also included.

Lakofsky, Charles. POTTERY. Art Horizons Series, Willard F. Wankelman, consulting ed. Dubuque, Iowa: William C. Brown Co., 1968. 113 p. 108 illus. Bibliography, p. 105. Glossary, pp. 107-10. Index, pp. 111-13. Paper.

Intended as neither a history nor a technical manual, this volume is a limited popular survey of the potter's art from the viewpoint of the nonpotter seeking a deeper understanding of its technicalities and creativity. Contemporary and historical examples are included in the illustrations.

Lane, Arthur. STYLE IN POTTERY. Park Ridge, N.J.: Noyes Press, 1974. 80 p. 49 illus. (1 color). Index, pp. 79-80.

Ignoring decorative tiles, porcelain figures, and similar types of work, Lane explains the term style as "the good handling of the potter's material." He examines a variety of vessels, which, through various ages and cultures as diverse as that of the ancient world and preindustrial Europe, he feels masterfully convey this sense of style.

Leach, Bernard. A POTTER'S PORTFOLIO: A SELECTION OF FINE POTS. London: Lund Humphries & Co., 1951. 88 p. Illus.

This oversize volume illustrates a collection of pots selected by the author on purely aesthetic grounds from all periods and backgrounds except that of industry. Containing sixty full page illustrations, this very handsome portfolio clearly reflects Leach's personal aesthetic convictions.

March, Benjamin. STANDARDS OF POTTERY DESCRIPTION. Occasional Contributions from the Museum of Anthropology of the University of Michigan, no. 3. Ann Arbor: University of Michigan Press, 1934. 55 p. 2 halftones, figs., charts. Paper. (Chamberlin, no. 1784).

This brief study attempts to formulate a form of description that treats all the major characteristics of pottery and can therefore be standardized and readily understood. March's descriptions of specimens are based on the four primary characteristics of ceramic material: paste, surface finish, decoration, and form.

Pickman, Dudley Leavitt. POURING VESSEL VAGARIES. Cambridge, Mass.: Mythology Co., 1938. 200 p. Chiefly illus.

This small picture book centers around the variety of pouring vessels that have come down to us through the ages. Pickman uses worldwide examples from the earliest pieces extant to those of the eighteenth century. While a few of these objects are formed from metal, the majority are clay pieces.

Rackham, Bernard. A KEY TO POTTERY & GLASS. New York: Chemical Publishing Co., 1941. xii, 180 p. 16 pls., figs. Index, pp. 173-80.

Intended as an informal history for the reader with some curiosity about the everyday articles around him. Rackham's study outlines the history of both pottery and glass from its origins in both the East and the West. As the emphasis here is on an understanding of the evolution of wares, Rackham has generally avoided the use of most specific names, dates, etc.

Rawson, Philip. CERAMICS: THE APPRECIATION OF THE ARTS/6. New York: Oxford University Press, 1971. 212 p. Illus. Index and glossary, pp. 207-12.

Although the author does include some historical as well as technical material in his presentation, his aim here is more clearly to explain the common basis of all ceramic history in terms of its human identity.

Rosenthal, Ernst. POTTERY & CERAMICS: FROM COMMON BRICK TO FINE CHINA. Harmondsworth, Middlesex, Eng.: Penguin Books, 1949. 304 p. 31 halftones, charts, tables. Glossary, pp. 292-301. Index, pp. 302-4.

Ranging from prehistoric pottery to modern industrial practices, this history covers the production of all types of ceramic work. Materials and processes are explained, and the properties of different types of clay work are also examined.

Savage, George. CERAMICS FOR THE COLLECTOR: AN INTRODUCTION TO POTTERY & PORCELAIN. New York: Macmillan Co., 1949. 224 p. 114 halftones. Glossary of oriental words, pp. 215-16. Index, pp. 217-24. (Chamberlin, no. 1757).

Intended as an introduction for the beginning collector, this volume briefly treats ceramic techniques and includes chapters on Far and Near Eastern work as well as on European pottery and porcelain.

*Savage, George, and Newman, Harold. AN ILLUSTRATED DICTIONARY OF CERAMICS. New York: Van Nostrand Reinhold Co., 1974. 320 p. 626 illus. (24 color).

This volume contains some 3,054 terms related to ceramics from antiquity to the present day. Although some technical terms are explained, the focus here is on historical material. Also included is an introductory list of the principal European factories and their marks.

Searle, Alfred B. AN ENCYCLOPEDIA OF THE CERAMIC INDUSTRIES.

See: Chapter 11: "Technical Works on Ceramic Materials and Processes" (p. 169).

[Slossom, Annie Trumbull]. THE CHINA HUNTERS' CLUB. New York: Harper & Brothers, Publishers, 1878. 274 p. 32 figs.

Narrated by the club's youngest member, the text describes several meetings of this group, whose members tell stories centering around some piece of pottery. Some technical or historical information on the pottery involved is always woven through the "folksy" tale. While its original readers may have found this approach quite charming, today it seems a silly example of nineteenth-century sentimentality. Slossom is the author's pen name.

Solon, M.L. [Louis Marc]. CERAMIC LITERATURE: AN ANALYTICAL INDEX.

See: Chapter 1: "Reference Works" (p. 4).

Sparkes, John C.L., and Gandy, Walter. POTTERS; THEIR ARTS & CRAFTS. London: S.W. Partridge & Co., n.d. 260 p. Figs. Index, pp. 257-60.

This old popular history surveys the techniques and history of ceramic art. Some basic information on clays and firing is included while individual chapters are devoted to ancient, medieval, oriental, and modern pottery.

Weiss, Gustav. THE BOOK OF PORCELAIN. Translated by Janet Seligman. New York and Washington, D.C.: Praeger Publishers, 1971. 335 p. Illus. (part color), tables. Bibliography, pp. 318-20. Index and glossary, pp. 324-35.

Weiss does not confine himself here to a traditional approach centering on factory names and marks. Rather he describes the development of porcelain within the context of political and cultural history, beginning with the origins of porcelain in China. Continuing along this line until the discovery of porcelain in Europe, he then describes the flowering of its manufacture there, the period of its decline, and its later revival. A survey of porcelain-producing centers as well as their marks is included in the appendixes. Although not too detailed, the text covers a substantial amount of technical information on porcelain bodies, production techniques, and the development of colors, machinery, and kilns.

William Rockhill Nelson Gallery of Art and Mary Atkins Museum of Fine Arts, Kansas City, Mo. RELIEF-ORNAMENTED CERAMICS: AN HISTORICAL SURVEY. Kansas City, Mo.: University Trustees, W.R. Nelson Trust, 1963. 32 p. Illus.

Organized in conjunction with the meetings of the Eighth Wedgwood International Seminar, this catalog offers a brief introduction to the use of applied, carved, and molded decoration by ceramic artists throughout the world. Of the fifty-six pieces exhibited, all but one or two are illustrated in the catalog.

Wykes-Joyce, Max. 7,000 YEARS OF POTTERY & PORCELAIN. New York: Philosophical Library, 1958. 276 p. 85 halftones. Bibliography, pp. 262-67. Index, pp. 268-76.

This popular history is a very informal survey of world ceramics from its beginnings to the twentieth century. Although it discusses both Far and Middle Eastern developments, the text leans most heavily toward European work.

Chapter 3

ANCIENT AND PRE-COLUMBIAN CERAMICS

(General Works, Eastern Hemisphere, Western Hemisphere)

GENERAL WORKS

Hartt, Ch. Fred. NOTES ON THE MANUFACTURE OF POTTERY AMONG SAVAGE RACES. Rio de Janeiro: Office of the S.A. Mail, 1875. 70 p. Paper.

The writer was the chief of the Geological Commission of Brazil, who, using evidence from "savage" tribes throughout the world, limited himself to considering the origins of ceramic art, the materials in use, the methods followed to complete a vessel and, particularly, the intimate connection between pottery and women among these peoples. Although most of Hartt's material is today simply curious, he does describe quite clearly the various hand-building techniques used by the different tribes in making their pottery.

Matson, Frederick R., ed. CERAMICS AND MAN. Chicago: Aldine Publishing Co., 1965. 301 p. Illus. Index, pp. 289-301.

This volume is one of a series of publications on research in general archaeology published by the Wenner-Gren Foundation for Anthropological Research. Comprised of eighteen papers delivered at a symposium in 1961 to evaluate the contributions of ceramic studies to archaeological and ethnological research, it is designed to be of service to the professional anthropologist. Each of the papers includes its own bibliography.

Shepard, Anna O[sler]. CERAMICS FOR THE ARCHAEOLOGIST. Carnegie Institution of Washington, Publication 609. Washington, D.C.: Carnegie Institution of Washington, 1954. xxxii, 414 p. 59 halftones and figs., 11 tables. Bibliography, pp. 395-405. Index, pp. 407-14. Paper.

Written from the viewpoint of a ceramic technologist, this volume examines the fundamental nature of ceramic materials and techniques. The author also seeks to explain in some detail the techniques of ceramic analysis and the evaluation of ceramic data.

EASTERN HEMISPHERE

Abercromby, John. A STUDY OF THE BRONZE AGE POTTERY OF GREAT BRITAIN AND IRELAND AND ITS ASSOCIATED GRAVE GOODS.

See: Chapter 5: "Western Ceramics: Greece to the Middle Ages" (p. 47).

*Amiran, Ruth. ANCIENT POTTERY OF THE HOLY LAND. New Brunswick, N.J.: Rutgers University Press, 1970. 360 p. 350 halftones, figs.

Including hundreds of photographs and diagrams throughout, this popular history covers in some detail the whole range of pottery making and use in this area from its beginnings in the Neolithic period (6000 B.C.) to the end of the Iron Age (600 B.C.). Chapters are ordered according to period and further divided by cultural group and pottery style.

Baumgartel, Elise J. THE CULTURES OF PREHISTORIC EGYPT. London: Oxford University Press, 1947. xi, 122 p. 13 halftones, 50 figs., 2 maps. 2 indexes, pp. 120-22.

While not dealing exclusively with the pottery of Egypt, more than half of this text greatly emphasizes pottery evidence to explain more clearly the foreign origins of the dynastic Egyptians. Baumgartel pursues the connection between the Egyptian decorated pottery of the Nakada II period and the painted wares of Iran and Iraq.

Benson, J.L. HORSE, BIRD AND MAN: THE ORIGINS OF GREEK PAINTING.

See: Chapter 5: "Western Ceramics: Greece to the Middle Ages" (p. 48).

Birch, Samuel. HISTORY OF ANCIENT POTTERY.

See: Chapter 5: "Western Ceramics: Greece to the Middle Ages" (p. 48).

Bjørn, Arne. EXPLORING FIRE AND CLAY: MAN, FIRE, AND CLAY THROUGH THE AGES. New York: Van Nostrand Reinhold Co., 1969. 88 p. Halftones, figs. Bibliography, p. 85. Index, pp. 86-88.

This small volume deals principally with contemporary experiments at the Archaeological Experimental Center in Lejre, Denmark, in Iron Age man's use of fire; about half of it is devoted to the methods of working with clay during that period. Of particular interest are the illustrations and experiments dealing with several Iron Age style kilns.

Clarke, D.L. BEAKER POTTERY OF GREAT BRITAIN AND IRELAND. 2 vols. Cambridge: At the University Press, 1970. xix, 576 p. 8 pls., figs., tables, maps. 7 appendixes, pp. 423-556. References, pp. 574-76.

Clarke's scholarly study aims at compiling an up-to-date corpus of material on the beaker in the form of scale drawings and measurements (over 1,000 are included in the text) and to reorganize the classification of material on the beaker (originally termed "drinking cup") so that it sensibly reflects the social groupings of its makers. Covering the period of British prehistory from 3000 to 1000 B.C., Clarke describes a system of British beaker classification, outlines European beaker origins, and breaks down his corpus into some fifteen British beaker groups.

Egami, Namio. THE BEGINNINGS OF JAPANESE ART.

See: Chapter 4: "Eastern Ceramics" (p. 38).

Furumark, Arne. THE CHRONOLOGY OF MYCENAEAN POTTERY. Stockholm: Kungl. Vitterhets Historie och Antikvitets Akademien, 1941. 156 p. 4 figs., tables. Bibliography, pp. 151-55. 2 indexes, pp. 132-50.

Forming the second part of a projected three-volume detailed study of Mycenaean pottery, Furumark's text presents here the external, purely archaeological, evidence for the chronological dating of the pottery. Both a relative chronology and an absolute chronology are presented.

______. THE MYCENAEAN POTTERY: ANALYSIS AND CLASSIFICATION. Stockholm: Kungl. Vitterhets Historie och Antikvitets Akademien, 1941. xix, 689 p. 71 figs. Bibliography, pp. 665-71. Index of sites, pp. 644-54. Index, pp. 672-78.

This heavily researched and enormously detailed piece of scholarship is the first of a projected three-volume study of Mycenaean pottery. THE CHRONOLOGY OF MYCENAEAN POTTERY (above) is part two of the study. A brief introduction to techniques and methods of analysis is included, but the great bulk of this text is devoted to analysis by shape and particularly decoration of the styles of Mycenaean pottery.

Gabus, Jean, and Junod, Roger Louis. AMLASH ART. Translated by Marjorie Gibson Craig. Orbis Pictus Series. Bern, Switz.: Hallwag, 1967. 9 p. 19 color pls.

The material here is from an exhibition by the Ethnographical Museum at Neuchatel, Switzerland. Included are a number of clay or bronze figures and vessels from the little known Amlash culture on the high Iranian plateaus.

HANIWA.

See: Chapter 4: "Eastern Ceramics" (p. 37).

Hayes, William C[hristopher]. GLAZED TILES FROM A PALACE OF RAMESSES II AT KANTIR. Metropolitan Museum of Art Papers, no. 3. New York: Metropolitan Museum of Art, 1937. 46 p. 13 pls. Bibliography, pp. 43-44. Index, pp. 45-46. Paper.

This footnoted paper examines the discovery, uses, and subject matter of the glazed tile fragments found or excavated at the palace site. The author feels that they represent an unequaled achievement of architectural ceramic glazed work in ancient Egypt.

Joseph, Adrian M., et al. CHINESE POTTERY BURIAL OBJECTS OF THE SUI AND T'ANG DYNASTIES.

See: Chapter 4: "Eastern Ceramics" (p. 32).

Kidder, J. Edward, Jr. THE JOMON POTTERY OF JAPAN. Ascona, Switz.: Artibus Asiae Publishers, 1957. 200 p. 7 halftones, 45 figs., 7 maps. Bibliography, pp. 154-86. Index, pp. 187-200.

Adapted from a dissertation submitted to New York University in 1953, and including a substantial bibliography, Kidder's text examines in detail the various types of Jomon pottery discovered in the different geographic areas of Japan. The figure illustrations compare pottery shapes to surface decoration while the halftone plates illustrate the surface decoration on fragments.

______. PREHISTORIC JAPANESE ARTS: JOMON POTTERY. Tokyo and Palo Alto, Calif.: Kodansha International, 1968. 308 p. 429 halftones, 16 color pls., charts. Bibliography and chapter notes, pp. 296-300. Index-glossary, pp. 301-8.

Principally composed of a number of excellent illustrations, this oversize and very handsome volume covers the whole range of Jomon pottery sites and periods. Included are a number of charts, lists, and maps indicating shape incidence, chronology, etc. Teruya Esaka contributed a chapter on decoration techniques and several of the charts.

Lacy, A.D. GREEK POTTERY OF THE BRONZE AGE. London: Methuen & Co., 1967; distributed in the U.S. by Barnes & Noble, New York. xv, 303 p. 4 color pls., 4 maps, 115 figs. and halftones. Bibliography, pp. 288-90. Index, pp. 291-303.

In this popular history Lacy is concerned with the 2,000-year period preceding the Iron Age (from 1100 B.C.). After a brief introduction to the Neolithic Era (6600-3000 B.C.) he devotes separate chapters to the Minoan Era (3000-1100 B.C.), the Helladic Era, the Mycenaean Era (1580-1100 B.C.), and the Cyclades and Cycladic pottery of the Bronze Age. The history and wares of each of these cultures are described in some detail.

Lapp, Paul W. PALESTINIAN CERAMIC CHRONOLOGY 200 B.C.-A.D. 70. Publications of the Jerusalem School, Archaeology, vol. 3. New Haven, Conn.: American School of Oriental Research, 1961. x, 231 p. Figs. 2 appendixes, pp. 221-31. Paper.

Originally submitted as a doctoral dissertation at the Harvard Divinity School, this extensively footnoted study attempts a clear presentation of Palestinian ceramic development by analyzing several categories of evidence and summarizing them in the form of a corpus of the pottery of the period. Economic, cultural, and historic implications of the evidence are briefly treated in the appendixes.

Levi, Doro. THE EARLY HELLENIC POTTERY OF CRETE. Princeton Monographs in Art and Archaeology, 23. Princeton, N.J.: Princeton University Press, 1945. 32 p. 105 halftones in 32 pls. Appendix, pp. 19-31. Bibliography, p. 32.

Originally published in HESPERIA, the journal of the American School of Classical Studies at Athens, this report briefly examines the transition of Cretan art from the dying off of Mycenaean history to the beginnings of Hellenic civilization in Crete.

Murray, Margaret A. CORPUS OF THE BRONZE-AGE POTTERY OF MALTA. London: B. Quaritch, 1934. 8 p. 46 pls., figs., maps.

Composing her description almost entirely of illustrations and diagrams showing the shapes and decorative designs on this pottery, the author briefly makes several conclusions concerning the archaeological excavations at the Hal Tarxien site on Malta.

Popham, Mervyn R. THE DESTRUCTION OF THE PALACE OF KNOSSOS. POTTERY OF THE LATE MINOAN IIIA PERIOD. Studies in Mediterranean Archaeology, vol. 12. Goeteborg, Swed.: P. Astroem, 1970. 111 p. 50 pls., figs. 2 appendixes, pp. 89-92. Paper.

Using selected deposits that belong to the period of the palace's final destruction or have a direct bearing on the date of its destruction, Popham attempts to define the character of the pottery in use at the palace and in the surrounding houses at the time they were destroyed. A number of examples and fragments are illustrated, and there are drawings of designs and contours.

Raphael, Max. PREHISTORIC POTTERY AND CIVILIZATION IN EGYPT. Translated by Norbert Guterman. The Bollingen Series, no. 8. [New York]: Pantheon Books, 1947. 160 p. 36 pls.

Raphael introduces his text with a description of the social and economic background of the era and devotes the remainder to discussions of the work of the basic cultural periods.

Schofield, J.F. PRIMITIVE POTTERY: AN INTRODUCTION TO SOUTH AFRICAN CERAMICS, PREHISTORIC AND PROTOHISTORIC. Handbook Series, no. 3. Capetown: South Africa Archaeological Society, 1948. 220 p. 14 pls., figs. Index, pp. 211-20. Paper.

> Schofield includes an introduction to pottery making among primitive peoples and identifies a variety of pottery from late Stone Age sites, Iron Age pottery, and more recent primitive pottery of South Africa. A good deal of ethnological as well as archaeological information is covered here.

Stubbings, Frank H. MYCENAEAN POTTERY FROM THE LEVANT. Cambridge: At the University Press, 1951. xvi, 111 p. 18 pls. 115 halftones, 37 figs., 3 maps.

> Originally completed for a doctoral thesis at Cambridge University, this revised study examines in detail Mycenaean pottery from Syria and Palestine, giving careful reference to its dating, distribution, and origin. In his conclusion, Stubbings uses this information to shed some light on Crete's position and influence in the Levant area.

Szentleleky, Tihamer. ANCIENT LAMPS. Translated by Arpad Debreceni; translation revised by D.M. Bailey. Chicago: Argonaut Publishers, 1969. 158 p. Illus. Indexes, pp. 149-58.

> This detailed work encompasses the ancient clay and bronze lamps from the fifth century B.C. to the sixth century A.D. which have been preserved in present day Hungary.

Taylour, Lord William. MYCENAEAN POTTERY IN ITALY, AND ADJACENT AREAS. Occasional Publications of the Cambridge University Museum of Archeology and Ethnology. Cambridge: At the University Press, 1958. 204 p. 17 pls., figs., maps. Index, pp. 197-204.

> While Taylour had originally hoped that his study would embrace the whole of the western Mediterranean the lack of the decisive ingredient--Mycenaean pottery--in outlying areas forced him to confine his work to the archaeological realities of evidence in the central Mediterranean, which was almost exclusively Italy. Taylour examines the evidence of Mycenaean influences in Sicily, the Aeolian Islands, Tarrato, Apulia, and a variety of other Italian sites.

Wallis, Henry. EGYPTIAN CERAMIC ART. London: Taylor, 1900. 37 p. 12 color pls., 45 figs.

> Beginning with a brief introductory text, this volume has as its major part a selection of watercolor or pastel illustrations drawn by the author. Many of the pieces illustrated are faience and each is accompanied by a detailed description. As is true of

Wallis's ART OF THE PRECURSORS (see p. 97), the age of its text and its illustrations' lack of photographic accuracy limit the usefulness of this study.

Welker, Marian. THE PAINTED POTTERY OF THE NEAR EAST IN THE SECOND MILLENNIUM B.C. AND ITS CHRONOLOGICAL BACKGROUND. Transactions of the American Philosophical Society, vol. 38, part 2. Philadelphia: American Philosophical Society, 1948. 62 p. 7 pls., figs., tables, map. Bibliography, pp. 229-33. Index, pp. 243-45. Paper.

Aimed primarily at a knowledgeable audience, Welker's paper analyzes the various painted wares which appeared in the region of the Khabur River and the Syrian coastal area between 2000 and 1200 B.C. The author goes into a good deal of detail concerning various styles and sites and their chronological sequence.

Wright, G. Ernest. "The Pottery of Palestine from the Earliest Times to the End of the Early Bronze Age." Photolithoprint. Ph.D. dissertation, Johns Hopkins University, 1937. 114 p. Figs., tables. Index, pp. 103-6.

Completed as a doctoral thesis, Wright's study explains the importance of pottery in the understanding and dating of archaeological material. The text is divided into three chapters, which examine the pottery of the Neolithic period, the Chalcolithic period, and the early Bronze Age. Several pages of notes, charts, and tables are also included.

Wu, G.D. [Chin-ting]. PREHISTORIC POTTERY IN CHINA. London: Kegan, Paul, Trench, Trubner & Co. The Courtauld Institute of Art, University of London, 1938. 180 p. 64 pls., figs., tables, maps. Bibliography, pp. 173-74. Index, pp. 175-80.

Originating as a thesis at the University of London, Wu's study incorporates all of the pottery-making periods before the Shang-Yin dynasty and examines these periods by area group, technical characteristics, and classes.

WESTERN HEMISPHERE

Boos, Frank H. THE CERAMIC SCULPTURES OF ANCIENT OAXACA. New York: A.S. Barnes & Co.; London: Thomas Yoseloff, 1966. 488 p. 4 color pls., 458 halftones, figs. Index, pp. 485-88.

Covering the period from 600 B.C. to 1350 A.D., this profusely illustrated text refers to a wide variety of carved and molded figures incorporated into urns, braziers, whistles, and plaques.

Breternitz, David A. AN APPRAISAL OF TREE-RING DATED POTTERY IN THE SOUTHWEST. Anthropological Papers of the University of Arizona, no. 10.

Tucson: University of Arizona Press, 1966. 128 p. Map, tables. 2 appendixes, pp. 109-18. References, pp. 119-28.

Using some 5,715 dated tree-ring specimens from 342 archaeological sites in the American southwest, this study dates 325 pottery types and varieties and ceramic categories from the introduction of fired ceramics to the time of the Spanish arrival (1550 A.D.).

Bushnell, G.H.S., and Digby, Adrian. ANCIENT AMERICAN POTTERY. Faber Monographs on Pottery and Porcelain. London: Faber and Faber, 1955. 51 p. 84 pls. (4 color). 117 illus., 2 maps, table. Bibliography, pp. 47-48. Index, pp. 49-51.

Although concise in its treatment, this volume does embrace a very wide range of material and includes a good number of illustrations. Organizing their subject by both locale and culture, the authors include information on the Anasazi, Hohokam, and Mogollin pottery cultures of the American southwest as well as those of Central and South America.

Izumi, Seiichi, and Tsunoyama, Yukihiro. TREASURES OF THE PRE-INCA CULTURES. Folio. Tokyo: San-Ichi Shobo, 1964; distributed by Maruzen Co., Tokyo. 56 p. Illus. (incl. 131 color pls., figs.). Bibliography, p. 56.

Printed in both Japanese and English, this lavish volume includes a five-page introduction to the prehistory of the Andean world by Yoshio Masuda and a detailed guide to the plates by the authors. Of the 131 tipped-in color plates, more than ninety show ceramic pieces while the remainder reproduce stone, metal, and fabric work.

*Lehmann, Henry. PRE-COLUMBIAN CERAMICS. Translated by Galway Kinnell. New York: Viking Press, 1962. 128 p. 32 pls. (8 color). Bibliography, pp. 122-25. Index, pp. 126-28.

Lehmann's popular study includes both a history of the ceramic cultures of pre-Columbian America and an explanation of their ceramic technology and principal ceramic forms, motifs, and decorative techniques.

Lilien, Rose. "A Study of Central Andean Ceramic Figurines." Ph.D. dissertation, Columbia University, 1956. 448 p. 23 pls., tables, figs. Bibliography, pp. 441-48.

This study aims at a comprehensive classification of central Andean figurines according to their types and subtypes. Covering the early formative cultures to the Imperial epoch, the thesis also compares Meso-American and central Andean figurines of the formative period. The text is supplemented by several pages of illustrations and numerous tables.

Lines, Jorge A. THE PRE-COLUMBIAN CERAMIC ART OF COSTA RICA. San Jose, Costa Rica: Instituto de Turismo, 1964. Unpaged. Illus., map. Bibliography. Paper.

Including information on pre-Columbian jewelry and sculpture as well as ceramics, this small pamphlet touches very briefly on the different styles of Costa Rican ceramics and offers a cursory popular history of that country's cultural heritage.

Longyear, John M. III. COPAN CERAMICS; A STUDY OF SOUTHWESTERN MAYA POTTERY. Carnegie Institution of Washington, Publication no. 597. Washington, D.C.: Carnegie Institution of Washington, 1952. xiii, 114 p. 118 pls., figs., tables, map. References, pp. 113-14. Paper.

The first part of this excavation study gives a general summary of the ceramic sequence at Copan and an account of Copan's history. The second part is devoted to a more detailed description of Copan pottery and artifacts.

Lothrop, S.K. TREASURES OF ANCIENT AMERICA: THE ARTS OF THE PRE-COLUMBIAN CIVILIZATIONS FROM MEXICO TO PERU. Geneva: Editions d'Art Albert Skira, 1964. 230 p. Illus. Index, pp. 227-29.

Dealing with carvings and metal objects as well as ceramic pieces, this large picture book very handsomely illustrates a number of ceramic masterpieces from pre-Columbian America.

Menzel, Dorothy, et al. THE PARACAS POTTERY OF ICA: A STUDY OF STYLE AND TIME. University of California Publications in American Archaeology and Ethnology, vol. 50. Berkeley and Los Angeles: University of California Press, 1964. 382 p. 15 pls., drawings, tables. Bibliography, pp. 305-7.

This volume is a general study of one of the major styles of Iça (Peru) pottery, based on associated and unassociated specimens.

Monti, Franco. PRE-COLUMBIAN TERRACOTTAS. Translated by Margaret Crosland. Cameo series. London: Paul Hamlyn, 1969. 158 p. 71 color pls.

Incorporating an extensive selection of illustrations, this popular survey describes the various techniques and different cultural styles of the terra-cotta ceramics produced in Central and South America. Each of the illustrations is identified and most are accompanied by a descriptive paragraph.

Patterson, Thomas C. PATTERN & PROCESS IN THE EARLY INTERMEDIATE PERIOD POTTERY OF THE CENTRAL COAST OF PERU. University of California Publications in Anthropology, vol. 3. Berkeley and Los Angeles: University of California Press, 1966. 180 p. 16 halftones, figs. Bibliography, pp. 133-36. Paper.

This detailed study examines thirteen consecutive phases of Mirimar and Lima pottery styles of the central coast of Peru when the peoples of this area possessed a distinctive art style devoid of foreign influence.

Proskouriakoff, Tatiana. GRAPHIC DESIGNS ON MESOAMERICAN POTTERY. Cambridge, Mass.: Peabody Museum; Washington, D.C.: Carnegie Institution, 1968. 19 p. Figs. Paper.

This brief monograph is by the author's own admission a cursory survey of graphic forms used on Mesoamerican pottery. Her emphasis is upon the ritual significance of decoration rather than the interpretation of decoration as "pleasing to the eye" or the product of individual expression.

Proulx, Donald A. LOCAL DIFFERENCES AND TIME DIFFERENCES IN NASCA POTTERY. University of California Publications in Anthropology, vol. 5. Berkeley and Los Angeles: University of California Press, 1968. 180 p. 64 halftones, figs., map, tables. Bibliography, pp. 118-20. Paper.

Originating as a graduate research project, this thesis seriates the early polychrome pottery from the Içа valley and examines the differences in shape and design between the pottery of the Nasca and the Iça valleys.

Rozaire, Charles E. ANCIENT CIVILIZATIONS OF LATIN AMERICA. Los Angeles: Los Angeles County Museum of Natural History, 1966. 72 p. Illus., map. Bibliography, p. 72. Paper.

This museum exhibition catalog includes a good selection of illustrations and a detailed introductory history to the dozen cultures exhibited. Although some stone and metal pieces are shown, the majority of the work is in clay.

Saville, Marshall H. A SCULPTURED VASE FROM GUATEMALA. Leaflets of the Museum of the American Indian, no. 1. New York: Heye Foundation, 1919. 5 p., folded insert. 2 figs.

Calling this bowl "the most beautiful example of earthenware ever found in either North or South America," Saville's publication describes the vase's sculptured relief decorations and reproduces them as a continuous flat panel in the folded insert.

Vaillant, George C. "The Chronological Significance of Mayan Ceramics." Ph.D. dissertation, Harvard University, 1927. vii, 392 p. 439 illus. in 58 pls. Tables, map. Bibliography, pp. 388-92. Archives of Archaeology, no. 12. Madison: Society of American Archaeology and University of Wisconsin Press, 1961. 7 microcards.

This extensive study remains a standard reference work in the archaeology of Mayan pottery. "Vaillant classifies pottery into 'ware' units, differentiating them by the technological attributes of paste and temper texture, vessel surface finish and over-all surface color as intended by the potters," states James C. Gifford in his foreword to the microcard edition, which in seven pages outlines the basic content and significance of Vaillant's text.

Weiant, Clarence Wolsey. AN INTRODUCTION TO THE CERAMICS OF TRES ZAPOTES, VERACRUZ, MEXICO. Bulletin of the Bureau of Ethnology, no. 139. Washington, D.C.: Smithsonian Institution, 1943. xiv, 144 p. 78 pls., 54 figs., 10 maps. Bibliography, pp. 133-37. Index, pp. 141-44. Paper.

Completed in partial fulfillment of a doctoral program, Weiant's report discusses a series of sampling excavations done in one season at a site encompassing a number of earth mounds in the Los Tuxtlas district. The various styles of pottery wares, figures, whistles, etc., are examined, and a concluding part of the study compares the different features of the pottery, relating them to Mexican cultural groups.

Chapter 4

EASTERN CERAMICS

(China, Korea, Japan, India, The Middle East)

General texts as well as volumes with overlapping subjects are included in the section on Chinese ceramics.

CHINESE CERAMICS

Barber, Edwin Atlee. HARD PASTE PORCELAIN (ORIENTAL). Art Primer, no. 9. Philadelphia: Pennsylvania Museum and School of Industrial Art, 1910. 46 p. 21 pls. Glossary, pp. 36-43. Index, pp. 45-46.

> Designed for the use of collectors and students of art history, this volume briefly examines the characteristics and styles of oriental hard-paste porcelain in China, Japan, Siam, and Korea. The last is covered in less than one page.

*Beurdeley, Cecile, and Beurdeley, Michel. A CONNOISSEUR'S GUIDE TO CHINESE CERAMICS. Translated by Katherine Watson. New York: Harper & Row, Publishers, 1974. 317 p. 111 color pls., 350 halftones and figs., maps. Appendixes, pp. 285-304. Bibliography, pp. 305-10. Index, pp. 313-17.

> Oversize and very handsomely illustrated, this beautiful text encompasses all the periods of Chinese ceramics from Neolithic times to the early twentieth century. Each chapter provides a chronological table, an introductory history of the period, and a catalog of examples using a number of small halftone illustrations with detailed captions. Potting techniques, styles, shapes, marks, and recent archaeological research are also treated.

Beurdeley, Michel. CHINESE TRADE PORCELAIN. 2d ed. Translated by Diana Imber. Rutland, Vt.: Charles E. Tuttle Co., 1963. 220 p. 344 illus., 24 color pls. Chronological table, pp. 210-11. Bibliography, pp. 212-13. Index, pp. 215-20.

> This well-illustrated study examines the forms and decoration of porcelains exported by the great charter companies of Europe, Russia, and the United States and examines the unique characteris-

tic of the wares intended for each country. Also included is an illustrated catalog of 250 outstanding examples.

Bulling, A[nneliese]. THE MEANING OF CHINA'S MOST ANCIENT ART. Leiden, Netherlands: E.J. Brill, 1952. 150 p. 8 pls., 79 figs. and halftones. 4 appendixes, pp. 136-42. Bibliography, pp. 143-48. Index, pp. 149-50.

Bulling's text interprets pottery patterns from Kansu (Ma Ch'ang and Pan-Shan) and their development in the Shang, Chou, and Han periods.

Donnelly, P.J. BLANC DE CHINE, THE PORCELAIN OF TEHUA IN FUKIEN. Monographs on Pottery and Porcelain. xiv, 407 p. 166 pls. (6 color). 5 appendixes, pp. 293-368. Index, pp. 369-94.

This thorough and handsomely illustrated volume covers in detail the entire history of blanc de chine ware, giving information on its types, dating, marks, and copies. Hundreds of pieces are illustrated and the illustrations are of generally good quality.

duBoulay, Anthony. CHINESE PORCELAIN. London: Octopus Books, 1963. 95 p. 32 color pls., 100 halftones.

Dividing his text into four sections--the early wares, the Ming dynasty, the Ch'ing dynasty, and Chinese export porcelains--duBoulay outlines the general history of these wares in a colorful picture book. The length of the text is not extensive and, while many subjects are mentioned, they are treated only briefly.

Gompertz, G[odfrey]. St. G[eorge].M[ontague]. CELADON WARES. New York and Washington, D.C.: Frederick A. Praeger, Publishers, 1969. 104 p. 40 pls. 52 illus. (4 color). 3 maps. Bibliography, pp. 98-99. Index, pp. 100-104.

Designed as an introductory history for the collector, this small volume defines the celadon of porcelain, explains its production, and discusses examples from China, Korea, Japan, and Siam (Sawankalok).

Gray, Basil. EARLY CHINESE POTTERY & PORCELAIN. Faber Monographs on Pottery and Porcelain, edited by W.B. Honey. London: Faber and Faber, 1953. 48 p. 120 halftones and 4 color illus. in 100 pls. Bibliography, p. 46. Index, pp. 47-48.

This volume examines the various styles and developments of Chinese ceramics from the Chou to the Yuan dynasties and illustrates a large selection of masterpieces from these periods.

Gulland, W[illiam].G[iuseppi]. CHINESE PORCELAIN. 2d ed. London: Chapman & Hall, 1899. 270 p. 495 illus. Pottery marks, pp. 245-63. Index, pp. 265-70. (Chamberlin, no. 1772).

Designed for the use of the amateur collector, this very old book covers a wide variety of information on porcelain. A good deal of space is devoted to explanations of the symbols and myths used in the ornamentation of these porcelains and to the classification of wares according to their glazing and coloring.

Hentze, C[arl]. CHINESE TOMB FIGURES: A STUDY IN THE BELIEFS AND FOLKLORE OF ANCIENT CHINA. London: Edward Goldston, 1928. 106 p. 114 pls. Bibliography, pp. 95-98. Index, pp. 99-106.

Originally published in French, this volume examines the origins and uses as well as the various styles and types of tomb figures. Although later research has no doubt dated some of this material, it offers a fairly detailed interpretation of these figures.

Hetherington, A[rthur].L[onsdale]. CHINESE CERAMIC GLAZES. London: At the University Press, 1937. 76 p. 14 color pls. Glossary, pp. 70-72. Bibliography, p. 73. Index, pp. 74-76. (Chamberlin, no. 1782).

Written for the collector, Hetherington's text attempts to briefly explain the scientific principles by which Chinese potters from the Han to the Manchu dynasties achieved their various glaze effects. Although a number of different effects are touched upon, the major portion of the text deals with the effects of using iron or copper.

_____. THE EARLY CERAMIC WARES OF CHINA. London: Benn Brothers, 1922. xviii, 160 p. 100 illus. (12 color) in 44 pls. Glossary of Chinese terms, pp. 150-53. Bibliography, pp. 155-56. Index, pp. 157-60.

Illustrating his text with what he considers to be typical examples, rather than extraordinary pieces, Hetherington informally describes for the collector the main characteristics of the products made in the Chinese factories before the Ming dynasty (pre-1368).

Hobson, R[obert].L[ockhart]. CHINESE ART. 2d rev. ed. Revised by Soame Jenyns. London: Ernest Benn, 1952. 21 p. 100 pls.

Although including examples of jade, bronze, lacquer ware, and painting, sixty-seven of the one hundred masterpieces selected by Hobson to form the core of this volume are ceramics dating from the Han to the Ch'ing dynasties.

_____. CHINESE POTTERY & PORCELAIN: AN ACCOUNT OF THE POTTER'S ART IN CHINA FROM PRIMITIVE TIMES TO THE PRESENT DAY. 2 vols. New York: Funk & Wagnalls Co.; London: Cassell & Co., 1915. 602 p. 136 pls. (40 color). Bibliography (vol. 1), pp. xxvii-xxx. Index (vol. 2), pp. 309-26. (Chamberlin, no. 1774).

Volume 1 of this fairly lengthy history covers the pottery and early wares of China while the second volume deals with a more detailed examination of the Ming and Ch'ing porcelains. A chapter in volume 1 is devoted entirely to identification, dating, and marks.

_____. THE LATER CERAMIC WARES OF CHINA. London: Ernest Benn, 1925. xxx, 156 p. 163 illus. (41 color) in 76 pls. Bibliography, p. xxiii. Index, pp. 151-56.

A sequel to Hobson's WARES OF THE MING DYNASTY and completing his trilogy, this volume covers the period of the Ch'ing dynasty of the Manchus (1644-1912). Most of the material deals with the more famous porcelain produced at Ching-te Chen, but Hobson has also included some information on pottery, stoneware, and porcelain made at other centers.

_____. THE WARES OF THE MING DYNASTY. London: Benn Brothers, 1923. xvi, 240 p. 128 illus. (11 color) in 59 pls. Bibliography, pp. 231-32. Index, pp. 235-40.

Designing this volume as a sequel to his EARLY CERAMIC WARES OF CHINA, Hobson examines here as many varieties of porcelain as possible from the Ming period. While pieces from several different sites are discussed, the majority are from the porcelain center at Ching-te Chen.

Hobson, R[obert].L[ockhart]., and Hetherington, A[rthur].L[onsdale]. THE ART OF THE CHINESE POTTER. London: Ernest Benn, 1923. xx, 20 p. 152 pls. (part color).

This large picture book illustrates and briefly describes 192 examples of Chinese ceramics from the Han dynasty to the end of the Ming dynasty. A brief introduction to the wares of each period is also included.

Hodgson, Mrs. Willoughby. HOW TO IDENTIFY OLD CHINESE PORCELAIN. Chicago: A.C. McClurg & Co., 1907. 178 p. 40 illus. Glossary, pp. 155-56. Bibliography, p. 157. Date marks and symbols, pp. 165-72. Index, pp. 173-76.

Intended as an aid to the amateur collector, Hodgson's book explains the dating of Chinese porcelains, gives a careful examination of their characteristics, and compares them to European examples. Although not actually a history, the book includes some historical information and also touches upon dynasties and periods, mythology, date marks, and symbols.

*Honey, William Bowyer. THE CERAMIC ART OF CHINA AND OTHER COUNTRIES OF THE FAR EAST. London: Faber and Faber and Hyperion Press, 1945. vii, 238 p. 195 pls. (3 color), map. 5 appendixes, pp. 190-217. Bibliography, pp. 218-26. Index, pp. 229-38. (Chamberlin, no. 1775).

Although principally devoted to the wares of China, Honey's text does embrace sections on Japan, Korea, and Indochina as well. The text is fairly thorough and includes an extensive number of notes and references. A large selection of clear, halftone illustrations supplement the text.

Hong Kong. University of. CHINESE TOMB POTTERY FIGURES. Hong Kong: 1953. 40 p. Illus. Bibliography, p. 15. Paper.

Written in both Chinese and English, this exhibition catalog includes some sixty-five illustrations, each with a brief identification, and an eight-page introduction in English by Jao Tsung-I.

Howard, David Sanctuary. CHINESE ARMORIAL PORCELAIN.

See: Chapter 6: "Western Ceramics: The Middle Ages to the Twentieth Century" (p. 74).

Hyde, J.A. Lloyd. ORIENTAL LOWESTOFT.

See: Chapter 7: "Ceramics of the United States, Canada, and Mexico" (p. 108).

Japan Society of New York. CHINESE, KOREAN AND JAPANESE POTTERIES. New York: 1914. 129 p. 66 illus. Glossary, pp. 55-57.

This descriptive catalog of an exhibition sponsored by the Japan Society includes a report on the ceramic wares of the Sung Dynasty by Rose Sickler Williams.

Jenyns, Soame. LATER CHINESE PORCELAIN; THE CH'ING DYNASTY (1644-1912). 4th ed. Faber Monographs on Pottery and Porcelain. London: Faber and Faber, 1971. xvi, 111 p. 228 illus. (8 color) in 128 pls. 3 appendixes, pp. 87-101. Bibliography, pp. 101-3. Index, pp. 105-11.

Since much work on this period had already been done by R.C. Hobson and W.B. Honey, Jenyns states that he only intends to update their work. He does, however, approach this study from a fresh viewpoint by centering his chronology around the three most famous directors of the porcelain factories at Ching-te Chen and by focusing his illustrations on pieces made for Chinese taste rather than for export. His final chapters discuss the period of decline (1750-1912) and the works of the provincial kilns.

______. MING POTTERY & PORCELAIN. Faber Monographs on Pottery and Porcelain, edited by W.B. Honey. London: Faber and Faber, 1953. 160 p. 249 illus. (4 color) in 120 pls. Bibliography, pp. 156-57. Index, pp. 158-60.

Containing a large number of illustrations, this detailed volume surveys the styles of porcelains produced during the reigns of the various Chinese emperors from the fourteenth to the seventeenth centuries A.D. A final chapter is devoted to export and provincial wares.

Joseph, Adrian M. MING PORCELAINS: THEIR ORIGINS AND DEVELOPMENT. London: Bibelot Publishers, 1971. 78 p. 109 halftones, tables, 1 color pl. 4 appendixes, pp. 35-41. Bibliography, p. 5.

Joseph here explains the rediscovery of early Ming porcelains and briefly describes, in separate sections, the materials and various decorative techniques used on Ming wares. He also presents a theory of placing the fifteenth-century blue and white wares chronologically within their period by the use of X-ray fluorescence spectrometry.

Joseph, Adrian M., et al. CHINESE POTTERY BURIAL OBJECTS OF THE SUI AND T'ANG DYNASTIES. London: Hugh M. Moss, 1970. 80 p. 118 halftones, map, tables. Bibliography, p. 79. Paper, price list insert.

This catalog of an exhibition arranged in conjunction with the Research Laboratory for Archaeology at Oxford University includes special reference to the scientific testing of pottery wares by thermoluminescence and to the works of the forger. Each of the 118 pieces is illustrated and described.

Koyama, Fujio. THE STORY OF OLD CHINESE CERAMICS. Translated by Jiro Harada. Tokyo: Mayuyama & Co., 1949. 34 p. 12 illus. in 5 pls., map. Paper.

Intended as an introduction and guide to Chinese ceramics for the beginner, this pamphlet clearly explains its stylistic growth through its various historic periods, beginning with the earliest painted pottery and continuing up to the Ch'ing dynasty. Although clearly written, the text sometimes covers styles of ceramic art in one or two paragraphs.

*Koyama, Fujio, and Figgess, John. TWO THOUSAND YEARS OF ORIENTAL CERAMICS. New York: Harry N. Abrams, n.d. 379 p. 160 illus. (part color). Bibliography, p. 379.

Oversize and handsomely illustrated, this popular history devotes several chapters to the wares of China and Japan, while those of Korea and Annam (Vietnam) are each treated in an individual chapter. Each of these subjects begins with a brief history of the country or period involved. The remainder of each section is devoted to large illustrations, which are each described clearly and in detail.

Lan P'u. CHING-TE-CHEN T'AO-LU OR THE POTTERIES OF CHINA. Translated by Geoffrey R. Sayer. London: Routledge and Kegan Paul, 1951. xxiv, 139 p. 2 appendixes, pp. 127-34. Index, pp. 135-39.

This volume is an eighteenth-century account by Lan P'u of the potteries at Ching-te-chen. Divided into ten books touching on a wide variety of ceramic information, its text examines the origins of the ware, procedures of production, and the reproduction of antique wares. With the T'ao Shuo, this account represents one of the few original documents of eastern ceramics available to the English-speaking reader.

Locsin, Leandro, and Locsin, Cecilia. ORIENTAL CERAMICS DISCOVERED IN THE PHILIPPINES. Rutland, Vt. and Tokyo: Charles E. Tuttle Co., 1967. xvii, 249 p. 221 illus. (part color). 3 appendixes, pp. 203-30. Glossary, pp. 231-32. Bibliography, pp. 233-35. Index, pp. 237-49.

This profusely illustrated and very readable volume records the archaeological work of the authors at the Santa Ana burial site in the Philippines. Their research sheds light on the various theories of Chinese trade wares and lends support to the position that blue-and-white pottery was made in south China as early as the Southern Sung dynasty. An interesting and valuable feature is the collection in the second appendix of small halftone illustrations showing the foot rims of some ninety-eight pieces illustrated throughout the text.

Los Angeles County Museum of Art. CERAMICS OF SOUTHEAST ASIA. Los Angeles: 1972. 12 p. 9 halftones. Paper.

This very small exhibition catalog briefly examines a selection of Khmer, Thai, and Annamese wares from Southeast Asia. Some historical information is included, although there is, as yet, relatively little known about them.

Mino, Yutaka. CERAMICS IN THE LIAO DYNASTY, NORTH AND SOUTH OF THE GREAT WALL. New York: China Institute of America, 1973. 86 p. 60 halftones, figs. Bibliography, pp. 20-26.

This exhibition catalog includes a nineteen-page introduction by the author and an extensive selection of annotated illustrations covering the various ceramic styles of the Liao dynasty.

Mudge, Jean McClure. CHINESE EXPORT PORCELAIN FOR THE AMERICAN TRADE.

See: Chapter 7: "Ceramics of the United States, Canada, and Mexico" (p. 109).

Okuda, Seeichi. ANNAMESE CERAMICS. [Tokyo]: n.p., 1954. 21 p. 96 illus. (6 color) in 72 pls.

Written in both Japanese and English and composed almost entirely of annotated illustrations, this volume briefly defines Annamese (Vietnamese) ceramics, outlines their history, and identifies the characteristics of the various wares.

Osgood, Cornelius. BLUE-AND-WHITE CHINESE PORCELAIN: A STUDY OF FORM. New York: Ronald Press Co., 1956. xvii, 166 p. 64 pls., figs., tables. Bibliography, pp. 147-55.

Anthropologist Osgood's study classifies these wares into ninety-two groups of form. Additional considerations of marks, motifs, and techniques of decoration are included.

Palmer, Bruce. "Pottery in the South Pacific." New York: Columbia University, Department of Art History and Archaeology, 1967. 21 p. 9 figs. Bibliography, pp. 18-21. Photocopy.

Presented in a symposium on early Chinese art and its possible influence in the Pacific basin, this paper (number 17) by the director of the Fiji Museum attempts to briefly summarize the elements needed to draw comparisons of pottery from the South Pacific with possible parent forms in southern China.

Penkala, Maria. FAR EASTERN CERAMICS: MARKS AND DECORATION. The Hague: Mouton & Co.; Rutland, Vt.: Charles E. Tuttle, 1963. 263 illus. Bibliography, pp. 235-41. Chinese, Korean, and Japanese glossaries, pp. 242-53. Index, pp. 254-63.

Encompassing China, Indochina, Korea, and Japan, this volume contains a brief history of ceramic development, as well as a fairly extensive examination of ornamentation and identification of markings, in each area.

Prodan, Mario. THE ART OF THE T'ANG POTTER. New York: Viking Press, 1960. 186 p. 34 color pls., 120 halftones, 6 diags. Bibliography, p. 179. Index, pp. 182-86.

This large volume contains a handsome selection of illustrations. Examples of T'ang pots as well as the more well-known burial figures of the period are represented.

Spelman, W.W.R. LOWESTOFT CHINA.

See: Chapter 6: "Western Ceramics: The Middle Ages to the Twentieth Century" (p. 84).

Spinks, Charles Nelson. THE CERAMIC WARES OF SIAM. Bangkok: Siam Society, 1965. 196 p. 54 pls. (2 color), figs., maps. Bibliography, pp. 177-96. Paper.

Spinks's study carefully examines the history of Thai ceramics, the various influences on its development, and the techniques of its production. Some fifteen pages of figure drawings illustrate the variety of traditional shapes and forms of these wares. Although it has some illustrations, this is more a volume of professional scholarship than a popular illustrated history.

______. SIAMESE POTTERY IN INDONESIA. Bangkok: Siam Society, 1959. 17 p. 65 pls. Paper.

Spinks briefly describes a variety of wares discovered in Indonesia and presently located in the Djakarta Museum. There are separate chapters devoted to Sukhothai, Chalieng, Sawankalok, and Annamese wares, as well as to Martaban and Cambodian jars. Although the illustrations are not so clear as they might be, they are adequate and do illustrate over one hundred examples of this pottery.

Staehelin, Walter A. THE BOOK OF PORCELAIN. Translated by Michael Bullock. New York: Macmillan Co., 1966. 87 p. 34 color pls.

This is an unusual popular treatment of the manufacture, transport, and sale of export porcelain during the eighteenth century, whose text has been developed around a series of thirty-four Chinese watercolors executed in the eighteenth century and discovered in 1954. Staehelin uses the contemporary texts of Pere d'Entrecolles and Pere du Halde, two French Jesuits, and of the Chinese potter T'ang Ying as a basis for describing the watercolors. Each of the watercolors has been reproduced here as a large tipped plate.

Thiel, Albert Willem Rudolf. CHINESE POTTERY AND STONEWARE. Los Angeles: Borden Publishing Co., 1953. 106 p. 120 pls. (24 color), 9 figs.

Containing a large selection of illustrations and excluding all porcelain, this volume touches on examples of different styles of Chinese pottery and stoneware from the Chou to the Ch'ing dynasties (1122 B.C. to 1912 A.D.).

Tokyo National Museum. EXHIBITION OF FAR EASTERN CERAMICS. Tokyo: 1970. 242 p. Illus. (16 color pls.). Paper.

Offering over 200 pages of photographs, this exhibition catalog covers Chinese, Korean, and Japanese ceramic masterpieces from the fifth century B.C. to the eighteenth century A.D. The foreword and the list of plates are in both Japanese and English. Unfortunately, "An Outline of Far Eastern Ceramics" and "A Chronological Chart of Far Eastern Ceramic History" are printed in Japanese only.

Valenstein, Suzanne G. MING PORCELAINS: A RETROSPECTIVE. New York: China Institute of America, 1971. 104 p. 70 halftones, 4 figs. Paper.

Valenstein describes the decorative changes and maturing skills of the Ming period from the fourteenth to the seventeenth centuries and attempts here to illustrate the broad scope of Ming wares, from the pieces considered of "imperial" quality to the relatively crude export ceramics.

Williamson, Dr. George C. THE BOOK OF FAMILLE ROSE. London: Kegan Paul, Trench, Trubner & Co., [1970]. xxiv, 231 p. 62 pls. (19 color). Bibliography, pp. xxi-xxiv. Index, pp. 215-31.

Intended as a guide to collectors and containing numerous illustrations, Williamson's text examines several features of the Yung Cheng and Ch'ien Lung porcelains. The author begins with a history of Chinese porcelains and includes letters sent by the early Jesuit missionaries on porcelain manufacture and early publications on Chinese porcelain in Europe and England. Additional chapters cover eggshell porcelain, famille rose decoration, classification, armorial china, figures, and further information for collectors. Each illustration is identified in some detail.

KOREAN CERAMICS

Akaboshi, Goro, and Nakamaru, Heiichiro. FIVE CENTURIES OF KOREAN CERAMICS: POTTERY AND PORCELAIN OF THE YI DYNASTY. Translated by June Silla. New York: Weatherhill; Tokyo: Tankosha, 1975. 159 p. 17 color pls., 112 halftones.

The short text of this volume contains several brief chapters dealing with Yi-Dynasty culture, its pottery, the tea ceremony, and the classification and terminology of its wares. The major portion of the book is filled with an extensive selection of clear illustrations of the wares and a description of each illustration is given at the end of the text.

Asia Society. THE ART OF THE KOREAN POTTER: SILLA, KORYO AND YI. New York: 1968; distributed by New York Graphic Society. 131 p. Illus., map. Bibliography, pp. 130-31.

This well-illustrated volume is the catalog of an exhibition of ceramics from the Silla, Koryo, and Yi dynasties, which Robert P. Griffing, Jr., selected for the Asia Society.

Gompertz, G[odfrey]. St. G[eorge].M[ontague]. KOREAN CELADON AND OTHER WARES OF THE KORYO PERIOD. Faber Monographs on Pottery and Porcelain. London: Faber and Faber, 1963. xvii, 102 p. 160 illus. in 108 pls. (6 color), map. Bibliography, pp. 91-96. Index, pp. 97-102.

Covering the period from 918 to 1392 A.D., Gompertz examines the outstanding achievements of the Koryo potter and the aesthetic significance of Korean celadon. A history of the Koryo wares, including their rediscovery in modern times, and the twelfth-century record of Hsu Ching are also given. The remainder of the text describes the types of wares and styles of decoration used during this period.

_____. KOREAN POTTERY AND PORCELAIN OF THE YI PERIOD. Faber Monographs on Pottery and Porcelain. London: Faber and Faber, 1968. xx, 106 p. 191 illus. in 128 pls. (8 color). Bibliography, pp. 87-94. Index, pp. 99-106.

Gompertz gives a brief history of the Yi period, extending from 1392 A.D. to the early twentieth century and describes the potter's craft during these times. Separate chapters are devoted to various wares and styles, including Punch'ong, white porcelain, underglaze iron and copper, blue-and-white, and black glazed wares.

Honey, W[illiam].B[owyer]. COREAN POTTERY. New York: D. Van Nostrand Co., 1948. 19 p. 100 pls. (4 color). Bibliography, p. 17. Index, pp. 18-19.

Although this publication is principally a collection of illustrations, Honey's introduction to it briefly examines the three basic period styles of Korean ceramics: the Silla, Koryo, and Yi.

*Kim, Chewon, and Kim, Won-Yong. TREASURES OF KOREAN ART: 2000 YEARS OF CERAMICS, SCULPTURE AND JEWELED ARTS. New York: Harry N. Abrams, 1966. 284 p. 300 illus. (24 color), maps, chronology. Glossary, pp. 257-60. Bibliography, pp. 261-63. Index, pp. 278-83.

This large, well-illustrated volume introduces the reader to the geographic and historic setting of Korean pottery and examines a variety of pieces from the Silla, Koryo, and Yi periods.

Mattusch, Harriett, and Mattusch, Kurt. ORNAMENTAL ROOF TILES OF KOREA. Seoul: Dong-A Publishing Co., 1962. 132 p. 61 halftones.

This is primarily a volume of illustrations, which consists of some fifty-two examples of Korean roof tiles from the Yi dynasty to the present. Each example is accompanied by interpretations of its symbolism. A brief history and explanation of the three basic styles of Korean roofs are also included.

JAPANESE CERAMICS

The Asia Society. HANIWA. New York: 1960. Unpaged. 49 illus., map.

This catalog of an exhibition of ancient clay figures from the graves of early Japan includes a very good selection of photographs of the pieces and a nine-page introduction to their history by Seiroku Noma.

Audsley, George A., and Bowes, James L. KERAMIC ART OF JAPAN. London: Henry Sotheran & Co., 1881. 304 p. 32 pls., marks. Index, pp. 289-304.

One of the earliest comprehensive studies of Japanese ceramics, this text includes a general essay on Japanese art and a history of Japanese ceramics treated by province. More emphasis is placed on the refined and elaborately decorated wares and the plates are executed in a harsh chromolithic technique. As with most pre-World War II books on Japanese ceramics, a good deal of this material is inaccurate.

Bowes, James L. JAPANESE POTTERY. Liverpool: Edward Howell, 1890. xxxi, 576 p. 16 pls., map. Index, pp. 557-76.

Bowes's text includes a brief history of Japanese pottery and an examination of the various Japanese kilns. The bulk of this volume, however, is devoted to the author's comments on examples from his own collection. A substantial number of pottery marks are also illustrated and explained.

_____. A VINDICATION OF THE DECORATED POTTERY OF JAPAN. Liverpool: Printed for private circulation, 1891. 63 p. 4 illus.

This is a lengthy reply to an anonymous review (by Edward S. Morse of Boston) criticizing Bowes's book THE KERAMIC ART OF JAPAN. Bowes had this rebuttal privately printed because he was unable to have it accepted by most of the public press. I think it reasonably accurate to describe Bowes's tone here as icily formal.

Egami, Namio. THE BEGINNINGS OF JAPANESE ART. Translated by John Bester. Heibonsha Survey of Japanese Art. New York: Weatherhill; Tokyo: Heibonsha, 1973. 178 p. 204 illus. (32 color), folded chart, map.

While some stone, bronze, and lapidary work is discussed in this popular history of Japanese art to the close of the Tumulus period (seventh century A.D.), the bulk of the text is about ceramics. The volume contains numerous illustrations of vessels and figures from the Jomon, Yayoi, and Tumulus periods and supplementary chapters by Teruya Esaka and Ken Amakasu on pre-Tumulus and Tumulus art, respectively.

Fujioka, Ryoichi. TEA CEREMONY UTENSILS. Translated and adapted by Louise Allison Cort. Arts of Japan, no. 3. New York: Weatherhill; Tokyo: Shibundo, 1973. 142 p. 204 illus. (part color). Glossary, pp. 135-37. Index, pp. 138-42.

Although nonceramic objects are also included, the majority of the pieces treated here are made of clay. Each of the ritual utensils is explained in a separate chapter. The text includes a brief introduction to the tea ceremony and an excellent selection of illustrations.

Fukui, Kikusaburo. HUMAN ELEMENTS IN CERAMIC ART. 2d ed. Tokyo: Society for International Cultural Relations, 1936. 16 p. 20 halftones. Paper.

This small pamphlet records a lecture given by Fukui to the Tokyo Women's Club in 1934 in which the author attempts to express some very personal ideas concerning his love and appreciation of ceramics.

Gorham, Hazel H. JAPANESE AND ORIENTAL CERAMICS. Rutland, Vt. and Tokyo: Charles E. Tuttle Co., 1971. 256 p. 36 pls. Bibliography, pp. 237-41. Index, pp. 243-56.

Intending to write an introductory text for the novice collector, Gorham touches upon a wide range of material dealing with Japanese ceramics. Included are historical outlines, clays, potters' methods, types of ware, kilns, influence of the tea ceremony, the use of designs and inscriptions, and advice on appreciation and purchasing. The author touches on Korean and Chinese ceramics, but both are discussed in relation to the history of Japanese ceramics.

Hisamatsu, Shin'Ichi. ZEN AND THE FINE ARTS. Translated by Gishin Tokiwa. Palo Alto, Calif.: Kodansha International, 1971. 400 p. 276 illus. (part color). Index, pp. 398-400.

Although this book illustrates examples from many areas of the fine arts, it has been listed here because it does reproduce a very handsome selection of Japanese ceramics and because it teaches an understanding of the spirit of Zen, which is an invaluable aid to the enjoyment of these works.

Iwamiya, Takeji. KATACHI: JAPANESE PATTERN AND DESIGN IN WOOD, PAPER AND CLAY. New York: Harry N. Abrams, [1963]. 170 p. 179 illus. (14 color).

With the exception of a brief introduction by Donald Richie, this entire volume is comprised of excellent photographs by Takeji Iwamiya showing examples of Japanese design from the eighth century to the present day.

Jenyns, Soame. JAPANESE PORCELAIN. London: Faber and Faber, 1965. xiii, 352 p. 272 illus. in 124 pls. (4 color), map. 5 appendixes, pp. 309-20. Bibliography, pp. 321-23. Index, pp. 325-52.

Jenyns describes the origins and development of porcelain manufacture in Japan, explaining its relationship to Japanese pottery as well as to the influences of China and the export trade. He covers the factories producing the arita, blue-and-white, kakiemon, kutani, and less important wares and in a final chapter relates several aspects of Chinese and Japanese porcelain manufacture.

_____. JAPANESE POTTERY. New York and Washington, D.C.: Praeger Publishers, 1971. 380 p. 120 pls. (4 color). 3 appendixes, pp. 314-47. Bibliography, pp. 348-50. Index, pp. 351-80.

This clearly written history covers the whole range of Japanese pottery from Neolithic times to the nineteenth century. Its detailed and thorough treatment makes this companion volume to Jenyns's JAPANESE PORCELAIN (above) most useful as a reference text.

Koyama, Fujio. THE HERITAGE OF JAPANESE CERAMICS. Translated by John Figgess. New York: Weatherhill, 1973. 252 p. 146 illus. (95 color). Appendix, pp. 241-47. Bibliography, p. 248. Index, pp. 249-52.

This beautifully illustrated and oversized book is divided into two parts. The several chapters of part 1 examine various important aspects of the Japanese ceramic tradition from its beginnings to the modern period while the second part is made up of a photographic essay documenting the continuing quality of that tradition today.

Leach, Bernard. KENZAN AND HIS TRADITION: THE LIVES OF KOETZU, SOTATSU, KORIN AND KENZAN. New York: Transatlantic Arts, 1966. 173 p. 106 pls. (12 color), drwgs., map. Index, pp. 169-73.

Leach attempts to portray the lives and work of these four masters in the context of the Genroku post-Renaissance near the close of the seventeenth century. He includes translated selections of diaries, poems, and two series of technical pottery notes.

Mikami, Tsigio. THE ART OF JAPANESE CERAMICS. Translated by Ann Herring. The Heibonsha Survey of Japanese Art. New York: Weatherhill; Tokyo: Heibonsha, 1972. 185 p. 200 illus. (44 color), maps.

Extensively illustrated, this very readable introduction to the history of Japanese ceramics covers the earlier and lesser known wares as well as those of the later periods.

Mitsuoka, Tadanari. CERAMIC ART OF JAPAN. Tourist Library, no. 8. [Tokyo]: Japan Travel Bureau, 1949. 190 p. 10 pls. (part color). Maps, pp. 173-83. Bibliography, p. 184. Index, pp. 185-90.

Although its bibliography is dated and the poor quality of its paper has obscured many of its illustrations, this small "tourist" book does include a good deal of information and gives a concise survey of Japanese pottery and porcelain. Several maps at the end help to give some visual perspective to the numerous kiln sites and place names cited.

*Munsterberg, Hugo. THE CERAMIC ART OF JAPAN: A HANDBOOK FOR COLLECTORS. Rutland, Vt. and Tokyo: Charles E. Tuttle Co., 1964. 272 p.

18 color pls., 200 halftones, map. Bibliography, pp. 265-66. Index, pp. 267-72.

Although its illustrations are sometimes not of the best quality, Munsterberg's profusely illustrated volume presents a brief updated survey of the periods and styles of Japanese ceramics from the prehistoric era to the present day.

Okuda, Seiichi, et al. JAPANESE CERAMICS. Tokyo: Toto Bunka Co., n.d. 222 p. 389 illus. in 106 pls. (10 color).

While only part of this volume is translated into English, that part does include a two-page outline of Japanese ceramics as well as a fairly detailed description of the work in each of the 106 plates.

THE RAKU-CHAWAN OF JAPAN. Tokyo: Akasaka Mitoko Co., n.d. 32 p. 32 illus. (3 color). Paper.

Written in both Japanese and English, this small pamphlet briefly describes the history of raku tea bowls and explains the appreciation of their aesthetics. A number of bowls are illustrated and the genealogy and marks of the raku family are also included.

Rhodes, Daniel. TAMBA POTTERY: THE TIMELESS ART OF A JAPANESE VILLAGE. Tokyo and Palo Alto, Calif.: Kodansha International, 1970. 180 p. 151 illus. (12 color), 2 figs. Bibliography, pp. 179-80.

This well-illustrated study examines in depth the lively, earthy ceramic forms of the Tamba region. Rhodes traces the history of the kilns at Tachikui, the last of the pottery villages in that area.

Sato, Masahiko. KYOTO CERAMICS. Translated by Anne Ono Towle and Usher P. Coolidge. Arts of Japan Series, no. 28. New York: Weatherhill; Tokyo: Shibundo, 1973. 133 p. 110 illus. Glossary, pp. 127-29. Index, pp. 131-33.

Written by a professor of Far Eastern art at Kyoto City University of Art, this well-illustrated volume covers the variety of Kyoto ceramics and examines the work of that area's two most famous potters, Ninsei and Kenzan.

Seattle Art Museum. CERAMIC ART OF JAPAN: ONE HUNDRED MASTERPIECES FROM JAPANESE COLLECTIONS. Seattle: 1972. 172 p. 132 illus. (32 color), maps. Glossary, pp. 161-67. Bibliography, pp. 169-72. Paper.

This is the catalog of an exhibition containing examples from both public and private collections and jointly sponsored by the Japanese government and several participating American museums. The catalog is well illustrated and includes a twenty-four-page introduction to the history of Japanese ceramics by Henry Trubner, curator of the department of Asiatic art at the Seattle Art Museum.

Stitt, Irene. JAPANESE CERAMICS OF THE LAST 100 YEARS. New York: Crown Publishers, 1974. ix, 214 p. 347 illus. (22 color). Appendix of marks, pp. 200-201. Glossary, pp. 202-6. Bibliography, pp. 207-8. Index, pp. 209-14.

Originally intending to cover only the export wares of the past one hundred years, Stitt somewhat expanded the scope of her text to include an introduction to the history and aesthetics of traditional Japanese ceramics and to the continuation of traditional styles in both the domestic and export wares of this period. The fairly detailed text surveys a very broad range of topics that include descriptions of old kilns, satsuma ware, the folk art movement, and those wares intended solely for the export market of the Nippon (1891-1921) and occupation (1945-52) periods.

200 YEARS OF JAPANESE PORCELAIN. St. Louis Art Museum and William Rockhill Nelson Gallery of Art and Mary Atkins Museum of Fine Arts. St. Louis: 1970. 190 p. 136 illus. (8 color). Bibliography, p. 190. Paper.

This well-illustrated exhibition catalog dealing with the development of Japanese porcelain from the second quarter of the seventeenth century to the early part of the nineteenth century has introductions by John A. Pope and Richard S. Cleveland.

Yanagi, Soetzu. THE UNKNOWN CRAFTSMAN: A JAPANESE INSIGHT INTO BEAUTY. Translated and adapted by Bernard Leach. Palo Alto, Calif.: Kodansha International, 1972. 232 p. 76 illus. (12 color). Index, pp. 225-30.

This selection of essays on oriental aesthetics, translated by the author's close friend, is by the father of the Japanese crafts movement, who gives his ideas on the Buddhist concept of beauty, the way of tea, and the responsibilities of the craftsman.

INDIAN CERAMICS

Manchanda, Mrs. Omi. A STUDY OF THE HARAPPAN POTTERY. Delhi, India: Oriental Publishers, 1972. 406 p. 178 figs., charts, map. Bibliography, pp. 381-86. Index, pp. 387-91.

This scholarly monograph examines examples of Harappan pottery in relation to the pre-Harappan and post-Harappan ceramic industries of the Indian subcontinent and contains a chapter on painted designs and art motifs. Also included are numerous figure illustrations of these decorative designs and profile-drawings of the different pottery shapes.

Starr, Richard F.S. INDUS VALLEY PAINTED POTTERY. Princeton, N.J.: Princeton University Press, 1941. xiii, 106 p. 174 figs., map. Index, pp. 103-5.

This revision of a doctoral dissertation submitted in 1938 examines the designs on the painted wares of the Harappan culture in northwestern India and compares the relationships existing between this culture and those of Iran and Mesopotamia.

MIDDLE EASTERN CERAMICS

Atil, Esin. CERAMICS FROM THE WORLD OF ISLAM. Freer Gallery of Art, Fiftieth Anniversary Exhibition, vol. III. Washington, D.C.: Smithsonian Institution, 1973. viii, 225 p. 200 illus. (part color), map, chart. 2 appendixes, pp. 219-24. Paper.

> This large catalog includes an introduction to the various styles of Islamic ceramics by the Freer's associate curator of Near Eastern art. Each of the 101 objects exhibited is illustrated, generally with an interior as well as an exterior view, and is described in detail.

Carswell, John. KUETAHYA TILES AND POTTERY FROM THE ARMENIAN CATHEDRAL OF ST. JAMES, JERUSALEM. 2 vols. Oxford: Clarendon Press, 1972. Vol. 1: THE PICTORIAL TILES AND OTHER VESSELS. xiv, 112 p. 46 pls. (part color), figs., map. 6 appendixes, pp. 102-12. Vol. 2: A HISTORICAL SURVEY OF THE KUTAHYA INDUSTRY AND A CATALOGUE OF THE DECORATIVE TILES. xiv, 106 p. 71 pls. (part color), 43 plans. 7 appendixes, pp. 59-96. Bibliography, pp. 97-98. Index, pp. 99-106.

> These two complementary but independent volumes supply a handsomely illustrated and comprehensive study of the Kuetahya pottery industry in northwestern Turkey. The first volume is devoted to a history of the pictorial tiles intended to repair the Holy Sepulcher in the eighteenth century but finally installed in the convent of St. James in Jerusalem. This part also includes a catalog of all inscribed and dated Kuetahya pottery from many different sources. Volume 2 traces the history of the Kuetahya pottery industry from its origins in the fifteenth century to the present day and contains an illustrated catalog of the different types of tiles among the thousands decorating the Armenian churches of Jerusalem.

Dimand, Maurice S. LOAN EXHIBITION OF CERAMIC ART OF THE NEAR EAST. New York: Metropolitan Museum of Art, 1931. 46 p. 36 halftones in 24 pls.

> Beginning with an introduction to the different styles of Near Eastern pottery, this catalog identifies and briefly describes each of the 203 pieces of Persian, Syrian, Egyptian, and Turkish pottery assembled for this exhibit.

Lane, Arthur. EARLY ISLAMIC POTTERY: MESOPOTAMIA, EGYPT AND PERSIA. 4th rev. impression. Faber Monographs on Pottery and Porcelain, edited by W.B. Honey and Arthur Lane. London: Faber and Faber, 1958. 52 p. 210 illus. in 100 pls. (4 color). Bibliography, p. 49. Index, pp. 51-52. (Chamberlin, no. 1776).

Attempting to show the cultural unity of early Islam in the relationship and development of its pottery wares from the eighth to the fourteenth centuries, Lane examines the wares of Rayy, Kashan, and Sultanabad. Also discussed are the earlier Mesopotamian wares of Bagdad and Samarra, the Samarkand pottery of central Asia, and the Rakka wares of northwestern Mesopotamia.

______. LATER ISLAMIC POTTERY: PERSIA, SYRIA, EGYPT, TURKEY. Faber Monographs on Pottery and Porcelain. London: Faber and Faber, 1971. 133 p. 191 illus. in 108 pls. (8 color), map. Appendix, pp. 119-23. Bibliography, pp. 124-27. Index, pp. 131-33. (Chamberlin, no. 1777).

Continuing from his EARLY ISLAMIC POTTERY (see preceding entry), Lane here discusses the later Persian, Mongol, blue-and-white, luster, and polychrome wares of the fourteenth to the eighteenth centuries. Separate chapters deal with Turkish pottery and Islamic pottery marks.

Mostafa, Mohamed. MOSLEM CERAMICS. Cairo: S.O.P. Press, 1956. 61 p. 95 illus. (10 color). Paper.

Written by the director of the Museum of Islamic Art, Cairo, this brief illustrated history describes the various periods (beginning with the eighth century) and styles of Islamic pottery and goes into some detail concerning the decorative techniques of the wares. While the illustrations generally are not of very good quality, they do encompass a wide range of examples that serve well enough as a brief survey of Islamic pottery.

Oz, Tahsin. TURKISH CERAMICS. [Ankara]: Turkish Press, Broadcasting and Tourist Department, n.d. 54 p. 75 pls. (39 color). Bibliography, pp. 47-49.

Covering material of the twelfth to the nineteenth centuries, this volume deals primarily with tile examples from various Turkish monuments. Several examples of Turkish pottery are also treated briefly.

Pier, Garrett Chatfield. POTTERY OF THE NEAR EAST. New York: G.P. Putnam's Sons, 1909. 175 p. 64 illus. Bibliography, p. xv.

This rather dated popular history of Middle Eastern ceramic wares briefly covers examples from Mesopotamia, Egypt, Syria, Turkey, and Persia.

Pope, Arthur Upham. AN INTRODUCTION TO PERSIAN ART SINCE THE SEVENTH CENTURY A.D. London: Peter Davies, 1930. xvi, 261 p. 103 halftones, maps. Bibliography, pp. 252-56. Index, pp. i-v.

This broad survey of Persian art has been included because of the limited amount of material dealing exclusively with Persian ceramics. Its author briefly explains the history of Persian pottery, its styles, and its use in architecture.

______. A SIGNED KASHAN MINA'I BOWL. Reprint Series, no. 7. New York: American Institute for Persian Art and Archaeology, 1936. 3 p. 1 color pl., 2 figs. Paper.

Reprinted from the October 1936 issue of THE BURLINGTON MAGAZINE, Pope's article describes the importance of the signature on this piece from the Kashan locale, identifying it with a great Kashan family of ceramic artists. Pope makes an interesting comparison here between the horse and rider painted on this bowl and a horse and rider silverpoint etching by Leonardo da Vinci. The intent is not a competitive evaluation of the two, but an instructive comparison of contrasting ideas and practices.

Rice, David Talbot. BYZANTINE GLAZED POTTERY. Oxford: Clarendon Press, 1930. xii, 120 p. 21 pls. (part color), figs., tables. Bibliography, pp. 105-7. Index, pp. 117-20.

Beginning with an explanation that very little is known of Byzantine pottery before the ninth century, Rice examines the varied form, decoration, and material of pottery ranging from the third to the fifteenth centuries. He also addresses the very complex question of origins in identifying these wares.

______. LATE BYZANTINE POTTERY AT DUNBARTON OAKS. Dunbarton Oaks Papers, no. 20. Washington, D.C.: Dunbarton Oaks Center for Byzantine Studies, 1966. Pp. 207-19. 14 halftone illus. in 12 pls.

This brief monograph of several pieces in the Dunbarton collection classifies these late Byzantine wares into five groups. The form, decoration, and techniques of each of the pieces are described in some detail.

Starr, Richard F.S. INDUS VALLEY PAINTED POTTERY.

See: "Indian Ceramics," above (p. 42).

Wallis, Henry. NOTES ON SOME EARLY PERSIAN LUSTRE VASES. Nos. 1, 2, 3. London: Bernard Quaritch, 1885-89. No. 1: 8 p. 9 figs., 1 color pl. No. 2: 11 p. 14 figs., 5 color pls. No. 3: 17 p. 20 figs., 7 color pls.

Wallis touches upon several aspects of early (thirteenth-century) Persian pottery, including its types of ware, places of origin, and forms of decoration. When these volumes were published, the study of these wares was very much in its infancy so that Wallis was able simply to comment on the examples illustrated, rather than to fit them into a clearly formulated history of the wares. The large but poorly colored plates appear to be reproductions of watercolor illustrations drawn by the author.

_____. THE ORIENTAL INFLUENCE ON THE CERAMIC ART OF THE ITALIAN RENAISSANCE.

See: Chapter 6: "Western Ceramics: The Middle Ages to the Twentieth Century" (p. 98).

Watson, H.O., & Co. AN ILLUSTRATED AND DESCRIPTIVE CATALOGUE OF RARE OLD PERSIAN POTTERY. New York: 1908. 81 p. Halftones.

Containing a brief history of Persian pottery and a description of its several types, this catalog lists some ninety pieces from a private collection being offered for sale by the Watson Galleries. While all of the pieces are described in some detail, not all of them are illustrated.

Wilkinson, Charles K. IRANIAN CERAMICS. New York: Asia Society, 1963; distributed by Harry N. Abrams, New York. 145 p. 99 pls. (part color), map. Glossary, p. 12. Bibliography, pp. 143-44.

This well-illustrated exhibition catalog surveys the history of Iranian ceramics from 4000 B.C. to the mid-nineteenth century and includes a historical introduction as well. Brief notes on each of the pieces in the exhibition are supplied and all but three are illustrated.

Chapter 5

WESTERN CERAMICS: GREECE TO THE MIDDLE AGES

While principally about Greek and Roman ceramics, a few volumes on very early European and English pottery are also included here.

Abercromby, John. A STUDY OF THE BRONZE AGE POTTERY OF GREAT BRITAIN AND IRELAND AND ITS ASSOCIATED GRAVE GOODS. 2 vols. Oxford: Clarendon Press, 1912. Vol. 1: 163 p. 61 pls. Vol. 2: 128 p. 49 pls. Index, vol. 2: pp. 125-28.

> Abercromby identifies the chronology of Bronze Age pottery in Britain and Ireland by zone or area as well as by class and type. Chapters are also included on colonization, continental influences, and comparative styles of ornamentation.

Arias, P[aolo].E[nrico]. A HISTORY OF 100 YEARS OF GREEK VASE PAINTING. Translated by B. Shefton. New York: Harry N. Abrams, [1962]. 410 p. 335 illus., 54 color pls. Index to collections, pp. 395-96. Index, pp. 397-410.

> The major part of this oversize volume is made up of large photographs illustrating the developing styles of vase painting. The text is coordinated to the illustrations and is organized by both style and artist.

Beazley, J[ohn].D[avidson]. ATTIC RED-FIGURE VASE-PAINTERS. 3 vols. 2d ed. Oxford: Clarendon Press, 1963. lvi, 2,036 p. 4 appendixes, pp. 1529-1616. 5 indexes, pp. 1707-2029.

> A revision and enlargement of editions published in 1925 and 1942. Beazley's catalog contains some 15,000 entries, including black-figure and white lekythoi work by red-figure painters. Ordering his material chronologically and including a selected bibliography for each painter, Beazley briefly identifies and describes the vases and fragments that can be attributed to each painter. Extensive appendixes and indexes identify provinces, collections, and so forth.

_____. ATTIC WHITE LEKYTHOI. The William Henry Charlton Memorial Lecture, November 1937. London: Oxford University Press, 1938. 26 p. 8 pls., 21 halftones. Bibliography. Paper.

Beazley briefly describes and illustrates the purpose, shape, techniques, and styles of the white lecythus, a vase type which appears to have lasted no more than one hundred years.

_____. THE DEVELOPMENT OF THE ATTIC BLACK FIGURE. 2d printing, corrected. Sather Classical Lectures, vol. 24. Berkeley and Los Angeles: University of California Press; London: At the University Press, 1964. 127 p. 49 pls. 2 indexes, pp. 123-27.

Consisting of the texts of several lectures, this volume presents the main lines of development of Attic black-figure vase decoration and discusses the styles of individual painters.

_____. ETRUSCAN VASE-PAINTING. Oxford Monographs on Classical Archaeology. Oxford: Clarendon Press, 1947. 352 p. 40 pls. Indexes, pp. 310-51.

Although centering on red-figure painting, this extensive work also includes studies of black-figure, superimposed color, black vases, silvered vases, and Latin pottery but does not treat the earliest products of the eighth and seventh centuries B.C.

_____. POTTER AND PAINTER IN ANCIENT ATHENS. Proceedings of the British Academy, vol. 30. London: Geoffrey Cumberlege Amen House, 1946. 43 p. 8 halftones. Paper.

This lecture, delivered to the Joint Meeting of Classical Societies at Oxford in 1942, touched upon a variety of subjects, including information derived from representations on vases, basic methods of production, signatures, and the relationship between potter and painter.

Benson, J[ack].L[eonard]. HORSE, BIRD AND MAN: THE ORIGINS OF GREEK PAINTING. Amherst: University of Massachusetts Press, 1970. 182 p. 41 pls., figs. 2 indexes, pp. 167-82.

Benson, a professor of art and archaeology at the University of Massachusetts, examines the style of the silhouette figures on Attic geometric vases, attempts to explain their origins, and places them in the history of pictorial expression.

Birch, Samuel. HISTORY OF ANCIENT POTTERY. Rev. ed. London: John Murray, 1873. xvi, 644 p. 209 figs. Appendix, pp. 600-614. Index, pp. 615-44.

Birch's broad text describing Assyrian, Egyptian, Greek, Etruscan, and Roman ceramic styles is supplemented by a chapter on Celtic, Teutonic, and Scandinavian pottery. However, the sections on

Greek and Roman pottery are the most extensive and clearly dominate the text. Although portions of it are undoubtedly very dated, there is a substantial amount of information on marks, forming techniques, kilns, and other technical material.

Boardman, John. ATHENIAN BLACK FIGURE VASES. New York: Oxford University Press, 1974. 252 p. 383 halftones. Bibliography, pp. 235-41. 3 indexes, pp. 249-52.

Boardman aims at presenting a detailed history of black-figure vases and includes enough illustrations to demonstrate the full range of their style. Although he omits explanations of potting and painting techniques, he includes works by both the finest and the less competent painters and discusses the shapes, chronology, and general decoration of these works and their depiction of scenes from reality and myth.

Brann, Eva T.H. LATE GEOMETRIC & PROTOATTIC POTTERY. The Athenian Agora, vol. 8. Princeton, N.J.: American School of Classical Studies at Athens, 1962. 134 p. 46 pls., figs., 2 maps. Bibliography, pp. xiii-xiv. Concordance, pp. 114-24. 2 indexes, pp. 125-34.

Referring to all of the late Geometric and proto-Attic pottery found in excavations by the American School before 1960, this detailed study treats its subject according to painting, shape, and topographical location.

Buschor, Ernst. GREEK VASE-PAINTING. Translated by G.C. Richards. London: Chatto & Windus, 1921. 180 p. 160 illus. Index of names, pp. 174-80.

Although this history covers the evolution of vase painting from the Stone and Bronze ages to the fifth century B.C., it emphasizes the red- and black-figure styles.

Cambitoglou, Alexander, and Trendall, A[rthur].D[ale]. APULIAN RED-FIGURED VASE-PAINTERS OF THE PLAIN STYLE. Monographs on Archaeology and Fine Arts, 10. N.p.: Archaeological Institute of America, 1961. 103 p. 41 pls. Indexes, pp. 87-103. Portfolio.

As its title indicates, this monograph deals with Apulian works executed in a plain rather than ornate style, classifying them by artist, more than fifty of whom are treated individually.

Caskey, L[acy].D[ale]. GEOMETRY OF GREEK VASES. Communications to the Trustees, 5. Boston: Museum of Fine Arts, 1922. 235 p. Illus. Paper.

Written by the museum's curator of classical antiquities as one of its occasional papers, this study applies the proportion discoveries of Jay Hambidge in his book DYNAMIC SYMMETRY: THE GREEK VASE (see p. 52) to some 185 vases in the museum's collection.

Coldstream, J[ack].N[ickolas]. GREEK GEOMETRIC POTTERY: A SURVEY OF TEN LOCAL STYLES AND THEIR CHRONOLOGY. Methuen's Handbooks on Archaeology. London: Methuen & Co., 1968. xxxix, 465 p. 64 pls., figs. Bibliography, pp. 391-94. Glossary, pp. 395-97. Indexes, pp. 399-456.

The author divides each local style according to its school, chronology, shape, and decoration. This large and detailed work of serious scholarship contains a good selection of illustrations (each plate combines five to twenty illustrations), a general index, an index of sites, and an index of collections.

Cook, R[obert].M[anuel]. GREEK PAINTED POTTERY. 2d ed. Methuen's Handbooks of Archaeology. London: Methuen & Co., 1972; distributed in the United States by Harper & Row Publishers, New York. 390 p. 56 pls., figs. Bibliography, pp. 331-59. Glossary, pp. 366-72. Index, pp. 373-90.

Written by a Lawrence professor of classical archaeology at Cambridge University, this study covers the schools of Greek painted pottery from the eleventh to the first century B.C., giving a brief account of black painted and relief wares. For the most part dealing with stylistic development, the text has additional chapters on the shapes, techniques, inscriptions, and history of the study of vase painting.

CORPUS VASORUM ANTIQUORUM. Union Academique Internationale. Paris: Edouard Champion, 1921-- . Halftones.

This very extensive collection of fascicles on classical vases has been published during the past fifty-five years in a format proposed and organized by the Union Academique Internationale and reported on in a small French volume entitled ORGANISATION DU CORPUS VASORUM ANTIQUORUM, published in 1921. Each fascicle is published by a particular museum, some museums having several fascicles. Belgium, Denmark, Germany, Spain, France, Great Britain, Greece, Italy, Low Countries, Poland, United States, and Yugoslavia each have a series. Since each of the fascicles is published more or less independently within the guidelines of the union, only some are in English.

Cox, Dorothy Hannah. THE EXCAVATIONS AT DURA-EUROPOS: THE GREEK AND ROMAN POTTERY. Part 1, fasc. 2. New Haven, Conn.: Yale University Press, 1949. 26 p. 5 pls., figs. Paper.

This is part of a final report made on the excavations conducted by Yale and the French Academy at Dura. The study covers the 600-year period of approximately 300 B.C. to 300 A.D., but according to Cox the excavations were disappointing because almost no wares were found intact. These wares included both imported and local types. Most of Cox's report is composed of figures illustrating the various shapes derived from the unearthed fragments.

Davison, Jean M. ATTIC GEOMETRIC WORKSHOPS. Rome: Bretschneider, L'Erma, 1968. 161 p. 64 pls., 3 appendixes, pp. 133-54.

Based upon a doctoral dissertation submitted in 1957, Davison's current study investigates the identity of "a new series of workshops composed of figured vases which seem closely associated by their decoration" and tries to discover any existing relationship between these workshops and both the Dipylon style and some late Geometric vases.

Debevoise, Neilson C. PARTHIAN POTTERY FROM SELEUCIA ON THE TIGRIS. University of Michigan Studies, Humanistic Series, vol. 32. Ann Arbor: University of Michigan Press, 1934. xiii, 132 p. 417 figs., 34 halftones in 14 pls., charts. Bibliography, pp. xii-xiii.

Beginning with a general history of the site excavations at Parthian levels, Debevoise centers his study on the excavations at Seleucia (Tel Umar). He briefly treats techniques of production, types of ware, glaze analyses, and chronology and fills most of the text with illustrations of the excavated shapes of the pottery.

DePuma, Richard D. ETRUSCAN AND VILLANOVAN POTTERY. Iowa City: University of Iowa Museum of Art, 1971. 35 p. Halftones, map. Bibliography, p. 3. Paper.

This catalog of an exhibition of some fifty-two pieces of Italian ceramics from Midwestern collections includes brief introductions to the five pottery styles forming the exhibit, a chronology of local historical events during 1500 to 100 B.C., and a fairly detailed description of each of the objects exhibited. Most of the pieces are illustrated.

Desborough, V.R. d'A. PROTOGEOMETRIC POTTERY. Oxford Monographs on Classical Archaeology, edited by J.D. Beazley and Paul Jacobsthal. Oxford: Clarendon Press, 1952. 330 p. 38 pls., map. 2 indexes, pp. 315-30.

Desborough examines the various ceramic styles in existence throughout the Greek world from the time of the latest Mycenaean style to the Geometric period, seeking to determine their interconnections.

Detsicas, A.P. THE ANONYMOUS CENTRAL GAULIST POTTER KNOWN AS X-3 AND HIS CONNECTIONS. Collection Latomus, vol. 64. Brussels: Latomus, Revue D'etudes Latines, 1963. 73 p. 15 pls., 184 figs. Bibliography, pp. 6-7. Paper.

Presenting a detailed study of this anonymous potter, commonly known in England as the "Anchor Potter," whose work is of particular importance in the study of Samian ware, Detsicas discusses the potter's dates, his early work, the site of his shop, his forms, and particularly the variety of decoration used on his wares.

Dyson, Stephen L. THE EXCAVATIONS AT DURA-EUROPOS: THE COMMONWARE POTTERY--THE BRITTLE WARE. Part 1, fasc. 3. New Haven, Conn.: Dura-Europos Publications, 1968. 94 p. 8 pls., figs. Index, p. 72. Paper.

Originally written as a doctoral thesis, this study is one of several completed as a record of the excavations at Dura. This part of the report deals with the utilitarian commonware always found in great abundance at Roman sites but often neglected in archaeological studies.

*Folsom, Robert S. HANDBOOK OF GREEK POTTERY: A GUIDE FOR AMATEURS. Greenwich, Conn.: New York Graphic Society, 1967. 213 p. 64 halftones in 16 pls., figs., charts. Bibliography, pp. 205-6. 2 indexes, pp. 207-13.

Covering the period 1050 to 146 B.C., this text emphasizes identification. Therefore, Folsom examines the clay itself; a repertory of designs; and the background, decoration, and shapes of pots. He also provides theoretical descriptions of pot making and decorating.

Frothingham, Alice Wilson. SIGILLATE POTTERY OF THE ROMAN EMPIRE. Hispanic Notes and Monographs. New York: Hispanic Society of America, 1937. xxx, 61 p. 124 illus. in 38 pls. Bibliography, pp. 51-55. Index, pp. 57-61.

This small volume describes the sigillate pottery (chiefly excavated in Spain by Arthur M. Huntington, president of the Hispanic Society, in 1937) that is in the society's collection. Including a brief history of sigillate pottery manufacture, centers of production, and marks, the major part of this publication is devoted to fairly detailed descriptions of each of the objects or fragments in the society's collection and to their illustration.

Hambidge, Jay. DYNAMIC SYMMETRY: THE GREEK VASE. New Haven, Conn.: Yale University Press; London: Oxford University Press, 1920. 161 p. 16 pls., diagrams.

Hambidge theorizes that there are two types of symmetry or proportion, active (dynamic) and passive. This study attempts to explain the rediscovery of a method in establishing the natural relationship of areas in design-composition (dynamic symmetry) that had been used by Greek artists in both their temples and their vases.

Hayes, J.W. LATE ROMAN POTTERY. London: British School at Rome, 1972. xxvii, 477 p. 55 halftones in 23 pls. 93 figs., 40 maps. 3 appendixes, pp. 428-50. Bibliography, pp. xix-xxv. Index, pp. 471-77.

The principal subject of this archaeological study are the fine tablewares--particularly the North African red-slipped wares and their imitations--produced in the Mediterranean provinces during the late Imperial period. Several groups of wares are identified and described and extensive catalogs of vessel types, decoration, stamps, and distribution are included in each chapter.

*Henle, Jane. GREEK MYTHS: A VASE PAINTER'S NOTEBOOK. Bloomington and London: Indiana University Press, 1973. 247 p. 79 halftones, figs. Bibliography, pp. 200-224. Index of types, pp. 225-47.

Using examples from the eighth to the fourth centuries B.C., Henle examines the representation of many of the Greek myths on vases. Some of the myths she discusses tell of the labors of Herakles, the Argonautica, the sack of Troy, and the Odyssey.

Herford, Mary A.B. A HANDBOOK OF GREEK VASE PAINTING. Manchester, Eng.: Manchester University Press, 1919. xxii, 125 p. 11 halftones, 21 figs. 3 indexes, pp. 117-25.

Intending her study for the nonspecialist, Herford gives a general introduction to the vases, examining their making, uses, and forms. The second half of her text covers the history of the vases from early examples to those done in Italy under Greek influence.

Higgins, R[eynolds].A[lleyne]. GREEK TERRACOTTAS. Methuen's Handbooks on Archaeology. London: Methuen & Co., 1967; distributed in the U.S. by Barnes and Noble, New York. liv, 169 p. 64 pls. (4 color), figs., map. Bibliography, pp. 135-57. Glossary, pp. 158-59. 3 indexes, pp. 161-69.

This illustrated study examines the statuettes and small reliefs of the Greek world from their earliest development around 7000 B.C. to the Hellenistic period (100 A.D.).

Hoffmann, Herbert. ATTIC RED-FIGURED RHYTA. Mainz, W. Ger.: P. von Zabern, [1962]. viii, 56 p. 24 pls. 3 indexes, pp. 53-56.

Hoffmann's study of these fifth-century Greek drinking cups (rhyta) chronologically divides them into groups so closely related in sculptural style that they are probably by the same potter or workshop. Hoffmann also discusses the work of the painters and includes illustrations of almost all his examples.

_____. TARENTINE RHYTA. Mainz, W. Ger.: Verlag Philipp von Zabern, 1966. viii, 167 p. 62 pls. 5 indexes, pp. 145-60.

In this sequel to the preceding reference, Hoffmann catalogs and describes some five hundred examples of these drinking vessels with animal-head bases. In addition to the various animal heads, he examines their relief decorations and provides clear illustrations of most of the pieces cataloged.

Hoppin, Joseph Clark. EUTHYMIDES AND HIS FELLOWS. Cambridge, Mass.: Harvard University Press, 1917. xvi, 186 p. 48 pls., figs. 4 indexes, pp. 177-86.

Hoppin presents a comprehensive study of several important vase painters who worked during the early period of the red-figured style. Principally concerned with the work of Euthymides, Hoppin discusses his signed vases, those attributed to him, and various aspects of his style. Hoppin also briefly examines the work of three of Euthymides' contemporaries--Phintias, Hypsis, and Kleophrades.

_____. A HANDBOOK OF ATTIC RED-FIGURED VASES. 2 vols. Cambridge, Mass.: Harvard University Press, 1919. Vol. 1: xxiv, 472 p. Halftones, figs. Vol. 2: xiii, 600 p. Halftones, figs. 4 indexes, pp. 501-600.

This extensive handbook briefly identifies the various vase painters of the sixth and fifth centuries B.C. About one hundred painters are identified and all their signed or attributed vases are described. Photographs of some of the vases are included, but most of the illustrations are of the painted panels and decorations on the vases.

_____. A HANDBOOK OF GREEK BLACK-FIGURED VASES. Paris: Edouard Champion, Editeur, 1924. xxiii, 509 p. Halftones, figs. 4 indexes, pp. 483-509.

In writing this companion volume to his earlier HANDBOOK OF ATTIC RED-FIGURED VASES (above), Hoppin intends to here include every signed vase from the seventh to the fourth centuries B.C. that can be called Greek and that had not been included in the earlier volume. Each of the resulting two hundred and fifty or so examples is illustrated, with the exception of about fifteen of which no photographs were available.

Huddilston, John Homer. LESSONS FROM GREEK POTTERY. New York and London: Macmillan Co., 1902. xiv, 144 p. 17 pls., figs. Bibliography, pp. 103-40. Index, pp. 141-44.

Believing that Greek vases are a source of understanding for many aspects of Greek life, Huddilston discusses history, religion, and the arts and relates the vases to several aspects of Greek daily life, including furniture, women, dress, war, and education. The illustrations are limited, but the bibliography, while very old, is extensive.

Johns, Catherine. ARRETINE AND SAMIAN POTTERY. London: British Museum, 1971. 31 p. 18 pls. (2 color), figs. Bibliography, pp. 28-30. Paper.

This small booklet briefly covers the history of these wares, dating them by describing their centers of manufacture, techniques, and changing of styles. A brief description accompanies each of the illustrations.

Karouzou, Semni. THE AMASIS PAINTER. Oxford: Clarendon Press, 1956. xii, 46 p. 44 pls. 3 appendixes, pp. 42-44. Bibliography, p. 45. Index, p. 46.

Karouzou describes the work of this potter and painter of the ripe black-figure style and illustrates a number of vases, particularly amphorae, which are either signed or can be attributed to him with some certainty.

*Lane, Arthur. GREEK POTTERY. 3d ed. Faber Monographs on Pottery and Porcelain. London: Faber and Faber, 1971. 64 p. 104 pls. (8 color). 179 illus. Bibliography, pp. 59-61. Index, pp. 63-64. (Chamberlin, no. 1768).

While Greek pottery has generally been the field of the archaeologist, Lane examines it in a wider sense and rejudges it on its merits as ceramic art, rather than interpreting it as an illustration of Greek life, thought, and literature. Methods of production, uses, styles, and a historical outline are included.

Mingazzini, Paolino. GREEK POTTERY PAINTING. Translated by F.B. Sear. Cameo series. London: Paul Hamlyn, 1969. 157 p. 69 color pls.

Containing a large number of excellent color plates, this popular survey describes the history of Greek vase painting, briefly explaining its various locales and styles from the twelfth to the fourth centuries B.C. Each illustration is identified and described.

Minoan Pottery.

See: Chapter 3: "Ancient and Pre-Columbian Ceramics".

Morgan, Charles H. II. THE BYZANTINE POTTERY. Corinth, vol. 11. Cambridge, Mass.: Harvard University Press, 1942. 373 p. 53 pls. (part color), figs. 3 appendixes, pp. 347-59. Index, pp. 363-73.

This report on some of the excavations carried out at Corinth by the American School of Classical Studies at Athens examines the development of Byzantine pottery and separately describes the plain glazed wares, the painted wares, and the sgraffito wares found there. Almost 2,000 examples are included in the catalog.

Mycenaean Pottery.

See: Chapter 3: "Ancient and Pre-Columbian Ceramics".

Myres, J[ohn].N[owell].L[inton]. ANGLO-SAXON POTTERY AND THE SETTLEMENT OF ENGLAND. Oxford: Clarendon Press, 1969. xvi, 259 p. 8 pls., Index, pp. 251-59.

Myres attempts to fill the gap in information about British history between Roman times and the period of Anglo-Saxon consolidation. He describes the types and distribution of evidence available and compares the pottery forms and decoration. While only a dozen or so pots are illustrated in the halftone plates, a wide variety of designs and decorative motifs are represented in the figures.

New York. The Metropolitan Museum of Art. SHAPES OF GREEK VASES. New York: 1922. 34 p. Halftones, figs. Paper.

With a minimum of text, this pamphlet illustrates and briefly describes about one hundred examples of Greek vases, cups, and jars. The examples are ordered according to shape and a few pages are devoted to the types of illustration decorating the surfaces of the vases.

Nicholson, Felicity. GREEK, ETRUSCAN AND ROMAN POTTERY AND SMALL TERRACOTTAS: A BRIEF GUIDE FOR THE SMALL COLLECTOR WITH A NOTE ON GREEK DRESS. London: Cory, Adams & Mackay, 1965. 67 p. 48 halftones, 20 figs. Bibliography, pp. 66-67.

Nicholson briefly describes the characteristic ceramic forms and their decoration for purposes of dating and identification. She includes an explanation of Greek dress and period hair styles as another aid to the collector.

*Noble, Joseph Veach. THE TECHNIQUES OF PAINTED ATTIC POTTERY. New York: Watson-Guptill Publications and Metropolitan Museum of Art, 1965. xvi, 218 p. 260 illus. (1 color), tables, graphs. Index, pp. 214-17.

Although including some material from other periods and cultures, the focus of this clear study is on the production of black-figure and red-figure Attic pottery between the seventh and fourth centuries B.C. Information on the dating, repair, and photographing of vases is given in the appendixes.

Oswald, Felix. THE TERRA SIGILLATA (SAMIAN WARE) OF MARGIDUNUM. Nottingham, Eng.: University College, 1948. 8 p. 53 pls., figs. Paper.

From his excavations at the site, Oswald identifies and briefly describes the Samian wares of some nine different periods from Margidunum's founding in 48 A.D. to the end of the fourth century.

Oswald, Felix, and Pryce, T. Davies. AN INTRODUCTION TO THE STUDY OF TERRA SIGILLATA. London: Longmans, Green & Co., 1920. 286 p. 85 pls., figs., map. Bibliography, pp. 245-72. Appendix, pp. 273-76. Index, pp. 277-86.

This comprehensive study of Roman terra sigillata places special emphasis on the chronological importance of the forms of these vessels and their decoration. The products of the Italian and the provincial (Gaulish) centers are examined and information on dates, stamps, forms, styles, and evolution is included.

Pereira, M. Helena Rocha. GREEK VASES IN PORTUGAL. Institute of Classical Studies. Portugal: University of Coimbra, 1962. 155 p. 46 pls. Bibliography, pp. 143-48. 3 indexes, pp. 151-55. Paper.

In a brief introductory essay, Pereira gives a historical account of the few vases excavated in Portugal and describes the formation of the vase collections acquired abroad and brought to Portugal during the nineteenth century. Some fifty pieces from Mycenaean to Roman times are identified and described in the catalog. Many are illustrated but the illustrations are often gray and cloudy.

Pfuhl, Ernst. MASTERPIECES OF GREEK DRAWING AND PAINTING. Translated by J.D. Beazley. London: Chatto & Windus, 1955. 152 p. 160 illus. (3 color). Bibliography, pp. 144-45. Index, pp. 148-52.

Pfuhl examines the various periods and styles of Greek painting from early Geometric through Etruscan wall painting to advanced Hellenistic. Almost two-thirds of the illustrations are of ceramic pieces.

Philippaki, Barbara. THE ATTIC STAMNOS. Oxford Monographs on Classical Archaeology, edited by J.D. Beazley et al. Oxford: At the Clarendon Press, 1967. 171 p. 64 pls., figs. 3 indexes, pp. 159-71.

Initially undertaken as a thesis for Oxford in 1947, this volume examines the shape, decorative scheme, and ornament of the stamnos form from its beginnings to the classic period.

Richter, Gisela M.A. ATTIC RED-FIGURED VASES: A SURVEY. New Haven, Conn.: Yale University Press; London: Oxford University Press, 1946. xxvii, 221 p. 125 illus. Bibliography, pp. vii-xii. Index, pp. 201-21.

Embracing the styles of red-figure vases between 530 and 320 B.C., Richter provides a brief explanation of their technical process and examines the work of many of the style's most prominent painters.

Richter, Gisela M.A., and Milne, Marjorie J. SHAPES AND NAMES OF ATHENIAN VASES. New York: Metropolitan Museum of Art, 1935. xxiii, 32 p. 191 halftones. Bibliography, pp. xvii-xxiii.

This small volume describes the periodic development of the various Greek vase forms and includes illustrations showing how each of the forms was originally used.

Robinson, Henry S. POTTERY OF THE ROMAN PERIOD CHRONOLOGY. The Athenia Agora, vol. 5. Princeton, N.J.: American School of Classical Studies at Athens, 1959. xiv, 149 p. 76 pls., figs. 2 indexes, pp. 141-49.

Dealing with some 850 specimens of pots and shards excavated from eight major deposits at the Athenian Agora, this text attempts to establish both a relative and an absolute chronology for the Roman wares of Athens used between the first century B.C. and the seventh century A.D.

Seeberg, Axel. CORINTHIAN KOMOS VASES. Institute of Classical Studies, Bulletin Supplement no. 27. [London]: University of London, 1971; distributed by International University Booksellers, London. xv, 107 p. 15 pls., chart. 3 indexes, pp. 83-107. Paper.

Attempting to explain the "padded dancer" symbol used by the Corinthian vase painters, Seeberg provides a nonselective survey of these figures on the known vases and fragments in this catalog of over two hundred examples. Many of these examples are illustrated.

*Sparkes, Brian A., and Talcott, Lucy. POTS AND PANS OF CLASSICAL ATHENS. Excavations of the Athenian Agora, Picture Book Number 1. Princeton, N.J.: American School of Classical Studies at Athens, 1958. 32 p. 62 halftones. Paper.

Taking a more informal and entertaining approach to classical Greek pottery than most texts, this small pamphlet illustrates a wide variety of ceramic household objects, from the handsome kylix to the most common cooking pot. The pieces illustrated are often coordinated with similar pieces in red-figure and black-figure vase paintings and the uses of many of these vessel types are described briefly.

Stanfield, J[oseph].A[loysius]., and Simpson, Grace. CENTRAL GAULISH POTTERS. University of Durham Publications. London and New York: Oxford University Press, 1958. liii, 409 p. 170 pls. (figs.), tables, map.

This extensive study begins with a history of Samian wares which describes the various pottery groups, workshops, types of wares, sites, and problems of identification. The bulk of the text is an extensive catalog of Trajanic and Hadrianic potters, each of whom is identified and whose examples of work are illustrated and described in some detail.

Stillwell, Agnes Newhall. THE POTTERS' QUARTER. Corinth; results of excavations, vol. 15, part 1. Princeton, N.J.: American School of Classical Studies at Athens, 1948. 138 p. 52 pls., figs., 3 maps. Index, pp. 135-38.

This first part of a report on excavations carried out at Corinth in the potters' quarter (or at least one of the potters' quarters) includes information on its buildings, architectural blocks, sculpture, metal and glass objects, and terra-cotta figurine moulds.

_____. THE POTTERS' QUARTER, THE TERRA-COTTAS. Corinth, results of excavations, vol. 15, part 2. Princeton, N.J.: American School of Classical Studies at Athens, 1952. 300 p. 59 pls., figs., map. Concordance, pp. 285-90. 2 indexes, pp. 291-300.

This is the second of a three-part report on the excavations at Corinth. Although this part does discuss terra-cotta lamps and miscellaneous clay and stone objects, about three-fourths of it is devoted to the terra-cotta figures discovered at the site.

Tiffany & Co. THE BARBERINI VASE: DISCOVERY, COMPOSITION AND DESCRIPTION OF PORTLAND'S MYSTIC URN.

See: Chapter 6: "Western Ceramics: The Middle Ages to the Twentieth Century" (p. 84).

Toll, Nicholas. THE EXCAVATIONS AT DURA-EUROPOS: THE GREEK GLAZED POTTERY. Part 1, fasc. 1. New Haven, Conn.: Yale University Press, 1943. 95 p. 20 pls. Paper.

This is part of a final report on a series of excavations completed at the town of Dura and its necropolis. The objects found are carefully examined according to material, technique, style, chronology, etc. The glazed lamps, figurines, altars, and sarcophagi that were also discovered are not discussed in this part of the report.

Trendall, A[rthur].D[ale]. THE RED-FIGURED VASES OF LUCANIA, CAMPANIA AND SICILY. 2 vols. Oxford Monographs on Classical Archaeology. Oxford: Clarendon Press, 1967. Vol. 1: xxix, 700 p. 1 color pl. Bibliography, pp. xxvii-xxix. Vol. 2: 112 p. 257 pls. 6 indexes, pp. 701-86.

This extensive study provides classified lists of Lucanian, Campanian, and Sicilian red-figure vases accompanied by explanatory comments on their style and dating. Volume 1 is divided into chapters dealing with individual painters and schools or vase groups of similar style. Each plate in volume 2 is comprised of approximately six illustrations.

_____. THE RED-FIGURED VASES OF LUCANIA, CAMPANIA AND SICILY. Supp. 1. Bulletin Supplement, no. 26. London: Institute of Classical Studies, 1970. xiv, 144 p. 28 pls. Index to collections, pp. 123-42.

This supplement attempts to bring up to date Trendall's original study of red-figure vases by including material from recent excavations and from previously unavailable sources.

_____. THE RED-FIGURED VASES OF LUCANIA, CAMPANIA AND SICILY. Supp. 2. Bulletin Supplement, no. 31. London: Institute of Classical Studies, 1973. xii, 294 p. 16 pls. Paper.

Trendall records an additional 400 Lucanian, Campanian, and Sicilian vases which have come to light since the publication of his first supplement in 1970.

Van Hoorn, G. CHOES AND ANTHESTERIA. Leiden, Netherlands: E.J. Brill, 1951. 200 p. 539 halftones in 144 pls.

Van Hoorn aims to explain the Athenian blossom feast of early spring, called the Anthesteria, by means of illustrating numerous examples of the choes, a squat-shaped oenochoe, made specifically for this festival. His catalog describes over 1,000 examples of these vases in European and American collections and includes examples of a similar type of vase made in southern Italy.

VonLuecken, Gottfried. GREEK VASE-PAINTINGS. 4 parts. The Hague: Martinus Nijhoff, 1923. 120 collotypes. Index. Paper.

Using a newly developed photographic process that allowed the paintings to be reproduced more naturally and without the distortions of the vases' curved and polished surfaces, VonLuecken illustrates approximately 120 vases from German collections. Except for his brief preface, there is no text.

Vos, M[aria].F[rederika]. SCYTHIAN ARCHERS IN ARCHAIC ATTIC VASE-PAINTING. Archaeologica Traiectina, 6. Groningen, Netherlands: J.B. Wolters, 1963. vii, 142 p. 20 pls. 2 appendixes, pp. 89-92. Bibliography, pp. 128-32. 2 indexes, pp. 133-41. Paper.

Vos divides his study into three periods--before 540 B.C., 530-490 B.C., and after 490 B.C.--and describes in some detail the use of Scythian archer figures as a design on Attic vases. Including over 400 examples of these vases, the catalog provides each with at least one bibliographical reference but illustrates only a limited number.

Walter, H[enry].B[eauchamp]. HISTORY OF ANCIENT POTTERY: GREEK ETRUSCAN AND ROMAN. 2 vols. New York: Charles Scribner's Sons, 1905. xxxvi, 1,092 p. 300 illus. (8 color). Bibliography, pp. xix-xxxvi. Index, pp. 557-88.

Although this work is probably of limited value today because of its age, it does present a fairly thorough treatment of Etruscan, Roman, and, especially, Greek pottery.

Webster, T[homas].B[ertram].L[onsdale]. POTTER AND PATRON IN CLASSICAL ATHENS. London: Methuen & Co., 1972. 312 p. 22 halftones in 16 pls. 3 indexes, pp. 301-12.

Limited by its lack of illustrations, this study by an emeritus professor of classics at Stanford University is basically divided into three parts: the identification of potters and painters; the various subjects of the vases; and the market for which these artists worked. Many examples are referred to simply by their identification in the works of Professor C.H.E. Haspels and Sir John Beazley.

Chapter 6

WESTERN CERAMICS: THE MIDDLE AGES TO THE TWENTIETH CENTURY

(English Pottery and Porcelain, European Pottery and Porcelain)

More general works that encompass material on English ceramics are included under European ceramics.

ENGLISH POTTERY AND PORCELAIN

Andrade, Cyril. ASTBURY FIGURES. London: 1924. 39 p. 31 halftones.

Composed principally of a collection of these early pottery figures exhibited at the Dalmeny Galleries, this catalog also briefly recounts the life and work of their maker, John Astbury (1678-1743). Since no ceramic work can be attributed to Astbury with absolute certainty, Andrade explains a number of characteristics which can aid in identification.

Barber, Edwin Atlee. ANGLO-AMERICAN POTTERY: OLD ENGLISH CHINA WITH AMERICAN VIEWS.

See: Chapter 7: "Ceramics of the United States, Canada, and Mexico" (p. 104).

Barker, P.A. THE MEDIEVAL POTTERY OF SHROPSHIRE FROM THE CONQUEST TO 1400. N.p.: Shropshire Archaeological Society, 1970. 80 p. 38 figs., 3 halftones. Appendix, pp. 44-45. Index, pp. 79-80. Paper.

This study is primarily concerned with the dating of material found at Shropshire sites and the economic pattern of the area rather than with the study of pottery as an art form. Nevertheless, a good deal of information on the pottery, including its decoration and uses, is given. Most of the figures illustrate shape outlines and comparisons of rim treatment.

Barnard, Julian. VICTORIAN CERAMIC TILES. London: Studio Vista Publishers, 1972. 184 p. 115 illus. Glossary, pp. 172-76. Bibliography, p. 179. Index, pp. 180-84.

Barnard examines the various makers, production methods, and numerous uses of ceramic tiles during the Victorian period. Writing chapters on industrial production in England, the American art tile companies, and William Morris and the craft tradition, he devotes a separate section to information on the collecting of tiles, which includes reproductions of the marks used by various manufacturers.

Barrett, Franklin A. WORCESTER PORCELAIN. Faber Monographs on Pottery and Porcelain, edited by W.B. Honey. London: Faber and Faber, 1953; New York: Pitman Publishing Corp., [1953]. 53 p. 147 illus. (4 color) in 100 pls. Marks, pp. 45-48. Bibliography, p. 49. Index, pp. 50-53.

Covering the development and various styles of porcelain made in Worcester, in western England, during the eighteenth century, Barrett also discusses the variety of influences, such as Chinese, Meissen, and Chelsea porcelains, on the ware and includes a good selection of illustrations.

Barrett, Franklin A., and Thorpe, Arthur L. DERBY PORCELAIN, 1750-1848. Faber Monographs on Pottery and Porcelain. London: Faber and Faber, 1971. xv, 206 p. 185 illus. in 104 pls. (8 color). 9 appendixes, pp. 71-198. Bibliography, pp. 199-200. Index, pp. 201-6.

Examining the period from 1750 to 1848, this history of one of the longest surviving porcelain companies in England discusses the work of Andrew Planche and treats separately the figures from the useful and decorative wares of the Chelsea-Derby period (1770-1784) and later. The appendixes provide lists of figures and information on factory workers, marks, pottery books, sale catalogs, and those pieces attributed to Planche.

Bedford, John. CHELSEA AND DERBY CHINA. Collectors' Pieces, no. 10. New York: Walker & Co., 1967. 64 p. Illus.

One of a series of small handbooks for the beginning collector, this very brief introduction to Chelsea and Derby china covers the history of these wares, their various styles, and the basic processes involved in their production.

_____. OLD WORCESTER CHINA. Collectors' Pieces, no. 9. New York: Walker & Co., 1966. 64 p. Illus.

This is another small handbook by Bedford on less expensive collectibles.

*Bemrose, Geoffrey. NINETEENTH CENTURY ENGLISH POTTERY AND PORCELAIN. Faber Monographs on Pottery and Porcelain. London: Faber and Faber, 1952. 57 p. 161 illus. in 100 pls. (4 color). Marks, pp. 44-50. Bibliography, pp. 51-52. Index, pp. 53-57.

Referring to a wide variety of wares produced in England during the nineteenth century, this study contains chapters on the country potteries, peasant wares, earthenware, Parian ware, transfer pottery, and the porcelains. There are also separate chapters on the Mintons and the great expositions (1840-70).

Beurdeley, Michel. CHINESE TRADE PORCELAIN.

See: Chapter 4: "Eastern Ceramics" (p. 27).

Binns, R.W. WORCESTER CHINA: A RECORD OF THE WORK OF FORTY-FIVE YEARS, 1852-1897. Edited by Charles F. Binns. London: Bernard Quaritch, 1897. xvii, 140 p. 33 pls. Appendix, pp. 115-35. Index, pp. 136-40.

Informally written as his professional memoirs by the director of the Royal Worcester factory. Binns's history describes the various types of wares that were developed during his tenure and recalls a number of important events, such as the great expositions of the late nineteenth century, which related to the work of the Worcester factory.

Blacker, J.F. THE ABC OF ENGLISH SALT-GLAZE STONEWARE FROM DWIGHT TO DOULTON. London: Stanley Paul & Co., 1922. 243 p. 200 illus. Index, pp. 239-43.

One of several "ABC" volumes by Blacker on English ceramic art, this work provides an account of the development in the seventeenth century, gradual decline, and eventual reemergence in the nineteenth century of salt-glaze ware in England.

_____. NINETEENTH CENTURY ENGLISH CERAMIC ART. Boston: Little, Brown & Co., 1911. 534 p. 1,200 figs. and halftones. Index, pp. 527-34.

Emphasizing late nineteenth-century ceramics, including the works of the Victorian period and the turn of the century, the text of this history provides somewhat difficult reading, and some elaborate examples singled out by the author for particular praise defy any description.

Bosanko, W. COLLECTING OLD LUSTRE WARE. The Collectors' Pocket Series. New York: George H. Doran Co., 1916. xv, 112 p. 46 halftones. Index, pp. 110-11.

Beginning with a brief introduction to lusterware and a history of its different types, Bosanko describes and illustrates a number of its examples produced at various English factories. Additional chapters are devoted to marks and makers, figures, and lustered porcelains.

Brears, Peter C.D. THE ENGLISH COUNTRY POTTERY: ITS HISTORY AND TECHNIQUE. Newton Abbot, Eng.: David & Charles, 1971. 266 p. 47 halftones and figs. Gazetteer, pp. 167-232. Appendixes, pp. 233-50. Bibliography, pp. 251-53. Index, pp. 255-66.

Attempting to record much of this material before it is lost, Brears explains the development of the country potteries from the medieval tradition, the urban potteries, art wares, and the typical processes of potting and firing. He also supplies indexes of potting terms and vessel names and a gazetteer of the sites, owners, and wares produced by the country manufacturers of England.

Bristol, Eng. City Art Gallery. BRISTOL PORCELAIN BICENTENARY EXHIBITION, 1770-1970. Bristol: 1970. 40 p. Illus. (part color). Bibliography, p. 6. Paper.

Commemorating the transfer of the Plymouth hard paste factory to Bristol, this exhibition catalog includes a very brief history by John Cooper and an explanation of the factory marks. Each of the 317 pieces exhibited is described and approximately twenty-five are illustrated.

Bristowe, W[illiam].S[yer]. VICTORIAN CHINA FAIRINGS. London: Adam & Charles Black, 1964. 115 p. 48 halftones.

"Fairings" refers to the small china figure groupings that were either given at or brought back from a fair. Covering the period of about 1860 to 1910, Bristowe explains that fairings, though conspicuously English, were actually made in Germany, one of the most important makers being in Elbogen. Bristowe brings the theme origins of these pieces into account and provides a catalog of the Elbogen pieces and a survey of other makers.

Burlington Fine Arts Club. EXHIBITION OF EARLY ENGLISH EARTHENWARE. London: 1914. xlvii, 150 p. 135 halftones in 50 pls.

This large, illustrated catalog surveys the history of English earthenware from the fourteenth to the eighteenth centuries. A forty-seven-page history of these wares by J.W.L. Glaisher introduces the catalog, whose remaining pages of text are devoted to fairly detailed descriptions of each of the pieces illustrated. Examples of salt-glazed ware, slipware, English delft, stoneware, animals, and figures are included.

Burton, William. JOSIAH WEDGWOOD AND HIS POTTERY. London: Cassell & Co., 1922. 195 p. 104 pls. (32 color). Index, pp. 185-95.

Dealing principally with the work of its founder, Burton examines the Staffordshire pottery industry and explains Wedgwood's influence on it. There are chapters on the various wares developed, Josiah Wedgwood's public work, his partners, family, principal artists, and, finally, the contemporary potters of Staffordshire.

Buten, Harry M., ed. WEDGWOOD TRIO BY METEYARD. Merion, Pa.: Buten Museum of Wedgwood, 1967. 280 p. 84 halftones, figs. Index, pp. 274-79.

This volume is a reprint of three books by Eliza Meteyard--WEDGWOOD AND HIS WORK (1873), MEMORIALS OF WEDGWOOD (1874), and CHOICE EXAMPLES OF WEDGWOOD ART (1879).

Church, Sir Arthur H. ENGLISH EARTHENWARE MADE DURING THE SEVENTEENTH AND EIGHTEENTH CENTURIES. London: Victoria and Albert Museum, 1911. xviii, 154 p. 78 pls., figs. Bibliography, pp. xvii-xviii. Index, pp. 151-54.

After several introductory chapters on the earliest English pottery, each further chapter of this handbook deals with a particular type of later ware, such as the early slipwares, Lambeth delftware, Eler's ware, Wedgwood, Staffordshire figures, etc. While the text is fairly detailed, most of the halftone illustrations are of later wares.

_____. JOSIAH WEDGWOOD: MASTER-POTTER. London: Seeley & Co.; New York: Macmillan & Co., 1894. 104 p. 4 pls., 28 halftones. Bibliography, p. 104.

Church's biography concentrates on the styles of work and developments contributed by Wedgwood to the ceramic industry. Wedgwood's jasperware, for example, is treated in several chapters.

Cook, Cyril. THE LIFE AND WORK OF ROBERT HANCOCK. London: Chapman & Hall, 1948. xvi, 88 p. 128 pls. Bibliography, pp. xiii-xvi. Index, pp. 84-88.

While the actual invention of English transfer printing can almost definitely be assigned to John Brooks, Hancock can be described as the creator of transfer printing because of his outstanding early work in this area. Cook explains Hancock's connection with the Battersea, Bow, and Worcester factories, and some 120 designs engraved by Hancock are illustrated and described in some detail.

Cooper, Ronald G. ENGLISH SLIPWARE DISHES 1650-1850. London: Alec Tiranti, 1968. 144 p. 249 illus. (part color) in 160 pls., 84 figs. Bibliography, p. 124. Index, pp. 138-44.

Cooper's readable and extensively illustrated text begins with a short history of world pottery and treats both the technique and history of English slipware. Slipware from the various regions of the British Isles, the work of Thomas Toft and other slipware potters, and a comparison of motifs and press-molded ware are discussed, but Cooper's emphasis is on the dishes, rather than the hollow or ornamental wares.

Coysh, A[rthur].W[ilfred]. BLUE-AND-WHITE TRANSFER WARE 1780-1840. Newton Abbot, Eng.: David & Charles, 1970. 112 p. 172 illus. (1 color). Bibliography, p. 9. Index, p. 112.

Offering a brief introduction to the technical process of creating transfer ware and a large selection of illustrations, this volume examines a number of the potteries which produced this ware and illustrates the variety of patterns employed by them.

CREAM-COLOURED EARTHENWARE. Victoria and Albert Museum, Small Picture Book, no. 53. London: Her Majesty's Stationery Office, 1960. 32 p. 38 halftones. Paper.

Principally composed of illustrations, this small pamphlet includes a three-page history of the cream-colored earthenwares.

Eaglestone, Arthur A., and Lockett, Terence A. THE ROCKINGHAM POTTERY. Rev. ed. Newton Abbot, Eng.: David & Charles, 1973. 159 p. 40 halftones in 16 pls. Marks, pp. 132-37. Bibliography, p. 146. Index, pp. 149-59.

Divided into two main parts, this volume first covers the entire history of the Rockingham pottery and explains the close potter-patron relationship that was so strong a part of that history; the second part discusses the Rockingham wares. This revised edition contains additional illustrations and an added chapter to bring it up to date with recent research, including the findings from site excavations conducted in 1967.

EARLY WEDGWOOD POTTERY. Barlaston, Stoke-on-Trent, Eng.: Josiah Wedgwood & Sons, 1951. xix, 110 p. Illus. (part color). Bibliography, p. ix. Glossary, p. 100-110. Paper.

Devoted chiefly to the wares produced by Josiah Wedgwood (1730-95), this small catalog covers his early pre-1769 wares as well as his basalts; jasper, marbled, and Queen's wares; and bone china. Several sections are devoted to his factories and to registry, trade, and date marks.

English Ceramic Circle. ENGLISH POTTERY AND PORCELAIN. London: Routledge & Kegan Paul, 1949. xv, 94 p. 122 pls. (3 color). Index, pp. 93-94.

Containing only a two-page introduction to the English Ceramic Circle as its text, this catalog of an exhibition sponsored by the circle illustrates and briefly identifies each of the 553 pieces in the exhibition. Grouped according to type or factory, the porcelains outnumber the earthenwares and stonewares by about four to one.

Finer, Ann, and Savage, George, eds. THE SELECTED LETTERS OF JOSIAH WEDGWOOD. London: Cory, Adams & Mackay, 1965. xi, 375 p. 15 pls. (2 color). Bibliography, pp. 345-46. Biographical index, pp. 347-70. Index, pp. 371-75.

To place his letters in some perspective, the editors preface their varied selection from some 1,300 of Wedgwood's extant letters with a historic outline of Wedgwood's life. Wedgwood's business expansion, sales, trade, production costs, catalogs, experiments, public affairs, children, politics, art, and science are all discussed in these letters, with the editors providing clarification where they think it is necessary.

Fisher, Stanley W. ENGLISH CERAMICS: EARTHENWARE, DELFT, STONEWARE, CREAM-WARE, PORCELAIN: INCLUDING A SECTION ON WELSH FACTORIES. New York: Hawthorn Books, 1965. 256 p. 178 halftones. Bibliography, pp. 47-49. Index, pp. 251-56.

While this volume does contain a brief history of English ceramics, it is principally composed of numerous illustrations and their identifying texts.

______. WORCESTER PORCELAIN. London: Ward Lock & Co., 1968. 96 p. 97 halftones. 4 appendixes, pp. 27-34. Index, pp. 95-96.

Primarily a picture book, Fisher's text includes an introductory history to the various periods of the Worcester factories from their beginnings in 1751 to the present day. Examples from all of the periods are illustrated and information on special services, patterns, gilders, and painters is included in the appendixes.

Gaunt, William, and Clayton-Stamm, M[axwell].D[avid].E[ugene]. WILLIAM DeMORGAN: PRE-RAPHAELITE CERAMICS. Greenwich, Conn.: New York Graphic Society, 1971. 176 p. 143 illus. (22 color). Bibliography, pp. 173-74. Index, pp. 175-76.

This examination of the life and work of the famous designer DeMorgan is supplemented by a large selection of excellent illustrations, which includes examples of his elaborate Persian designs and brilliant luster decoration.

Godden, Geoffrey A. BRITISH PORCELAIN: AN ILLUSTRATED GUIDE. New York: Clarkson N. Potter, Publisher, 1974; distributed by Crown Publishers, New York. 451 p. 12 color pls., 578 halftones. Bibliography, pp. 446-48. Index, pp. 449-51.

Intending to complement his ILLUSTRATED ENCYCLOPEDIA OF BRITISH POTTERY AND PORCELAIN with this volume composed almost entirely of illustrations, Godden has selected these illustrations to record key shapes and to illustrate the wares of the lesser-known factories that frequently have no literature of their own.

A brief chronological survey is included and the illustrations are alphabetically arranged by manufacturer.

______. BRITISH POTTERY AND PORCELAIN, 1780-1850. London: Arthur Barker, 1963. 199 p. 67 halftones in 40 pls., marks. Appendix, p. 185. Bibliography, pp. 186-89. Index, pp. 190-99.

Godden touches briefly upon the whole range of British pottery created during this period. Although the larger and more well known potteries, such as Wedgwood and Staffordshire, are cited, Godden has deliberately treated them quite briefly so that he has been able to give more space to neglected manufacturers, special bodies, and the northern Welsh and Scottish factories. Chapters are also devoted to marks, printing, reproductions, and fakes.

______. CAUGHLEY AND WORCESTER PORCELAINS, 1775-1800. New York: Frederick A. Praeger, Publishers, 1969. xxii, 166 p. 317 halftones in 170 pls. (10 color), map. 3 appendixes, pp. 151-61. Bibliography, p. 163. Index, pp. 164-66.

In writing this introductory history of Caughley porcelain and a description of its marks and designs, Godden pays particular attention to the study of the factory wasters that had been found during the previous few years at the Caughley site. Godden explains that the wasters show that many wares attributed to Caughley were actually by Worcester and that in many cases the Caughley factory did not copy the Worcester pieces, as was previously believed. Numerous examples from both factories are illustrated, often with matching wasters from the Caughley site.

______. COALPORT & COALBROOKDALE PORCELAINS. New York: Praeger Publishers, 1970. xv, 156 p. 247 illus. in 154 pls. (10 color). 2 appendixes, pp. 143-51. Bibliography, p. 152. Index, pp. 153-56.

Beginning with a general history and including information on marks, individual artists, and factory excavations, Godden identifies and illustrates (ninety plates) the early unmarked porcelain that had previously been attributed to other English factories, particularly Worcester. The later wares of the Coalport factory are also treated and pages from several early pattern and style books are included in the illustrations.

______. THE ILLUSTRATED GUIDE TO LOWESTOFT PORCELAIN. The Illustrated Guides to Pottery and Porcelain. New York: Frederick A. Praeger, Publishers, 1969. xix, 164 p. 225 illus. in 112 pls. (8 color). 2 appendixes, pp. 127-56. Bibliography, pp. 157-59. Index, pp. 162-64.

The first of a projected illustrated guide series for collectors, this extensively illustrated volume includes a general history of the Lowestoft factory as well as information on marks, dating, and forms. Fakes and false attributions are treated in the second appendix.

______. JEWITT'S CERAMIC ART OF GREAT BRITAIN 1800-1900. New York: Arco Publishing Co., 1972. xxviii, 282 p. 126 illus. (10 color), 54 engravings. 2 appendixes, pp. 257-62. Glossary, pp. xvi-xxviii. Bibliography, pp. 263-66. Index, pp. 267-82.

This volume is a revised edition of parts of Llewellynn Jewitt's THE CERAMIC ART OF GREAT BRITAIN FROM PRE-HISTORIC TIMES DOWN TO THE PRESENT DAY, which was first published in 1878. Godden has updated the most valuable parts of Jewitt's work, his discussion of the nineteenth century, where he recounted a good deal of firsthand information on the histories of hundreds of smaller manufacturers. About half of the text is devoted to the Staffordshire potteries and information on marks and dating as well as errors in the 1883 edition are noted in the appendixes.

______. MINTON POTTERY AND PORCELAIN OF THE FIRST PERIOD 1793-1850. London: Herbert Jenkins, 1968. xvi, 168 p. 162 illus. in 94 pls. (12 color). 3 appendixes, pp. 141-63. Bibliography, p. 164. Index, pp. 165-68.

Inspired by his discovery of early pattern books in the Minton factory archives, Godden identifies numerous pieces of Minton porcelain from the "lost period," which had previously been attributed to Coalport, Derby, Spode, Worcester, and others. Many of these patterns are illustrated here with the actual Minton products. In addition to chapters on tablewares, figures, and ornamental pieces, the author also includes information on designers, artists, modelers, and marks.

______. VICTORIAN PORCELAIN. The Victorian Collector Series. New York: Universe Books, 1970. 222 p. 102 illus. 2 appendixes, pp. 212-14. Bibliography, pp. 215-16. Index, pp. 217-22.

This general survey of wares produced by the English porcelain factories during the Victorian era makes extensive use of contemporary sources and places special emphasis on the ceramic artists of the period. Lists of artists employed by many of the principal factories are included as are identifications of numerous factory marks.

Gosse, Edmund. SIR HENRY DOULTON. Edited by Desmond Eyles. London: Hutchinson & Co., 1970. xiv, 218 p. 2 appendixes, pp. 189-211. Index, pp. 213-18.

This biography of the great Victorian potter was written in 1899-1900 by Sir Edmund Gosse but was never published by the Doulton family. Only with the death of Eric Hooper, Doulton's grandson, was permission granted for publication. The appendixes contain several addresses and extracts of lectures by Sir Henry himself. There are no illustrations.

Haggar, Reginald G. ENGLISH COUNTRY POTTERY. London: Phoenix House, 1950. 160 p. 64 halftones in 32 pls., figs. Marks, pp. 138-44. Bibliography, pp. 145-53. Glossary, pp. 154-55. Index, pp. 156-60.

Intending his book as a guide for the collector, Haggar roughly divides the English earthenware tradition into preindustrial and postindustrial periods with a chapter on Josiah Wedgwood between the two. The preindustrial period includes slipware, salt-glazed ware, and chimney ornaments while the postindustrial period covers the cream-colored wares, transfer printing, and Victorian and modern pottery.

______. STAFFORDSHIRE CHIMNEY ORNAMENTS. New York: Pitman Publishing Corp., 1955. 158 p. 5 color pls., 98 halftones, 21 figs., marks. 6 appendixes, pp. 115-50. Bibliography, pp. 151-52. Index, pp. 153-58.

The author examines the work of some thirty-six different individuals or families producing pottery figures in England from the end of the seventeenth century to the latter part of the nineteenth century. According to the author, this text is designed to fill the gaps left by Herbert Read's STAFFORDSHIRE POTTERY FIGURES, published in 1929.

Hayden, Arthur. CHATS ON OLD EARTHENWARE. New York: Frederick A. Stokes Co., 1909. 496 p. 150 halftones, marks. Bibliography, pp. 23-25. Glossary, pp. 27-32. Index, pp. 485-96.

Intending his text as a guide for the modest collector of English earthenware, Hayden avoids reference to rare museum examples. The early wares, English delft, stoneware, Staffordshire ware, salt-glazed ware, Wedgwood, Leeds, transfer ware, figures, and lusterware are each treated separately and the author includes advice for the collector and lists of recent prices.

______. SPODE AND HIS SUCCESSORS: A HISTORY OF THE POTTERY STOKE-ON-TRENT 1765-1865. London: Cassell & Co., 1925. xxiii, 204 p. 88 pls. (24 color). Bibliography, p. xix. Index, pp. 197-204.

This history of the Spode family and factory describes the contributions made by Josiah Spode, particularly his introduction of blue underglaze transfer printing and bone china, and by Josiah Spode II, who introduced feldspar porcelain and stone china. Hayden continues the history into the Copeland period, stopping at 1865. Spode pieces are illustrated and the author also shows designs from the original copper plates used in transfer printing, a number of original molds, and various factory documents and records.

Hobson, R[obert].L[ockhart]. MEMORIAL LECTURE ON MR. M.L.E. SOLON. [Stoke-on-Trent: English Ceramic Society, 1914-15.] From the TRANSACTIONS OF THE ENGLISH CERAMIC SOCIETY, 1914-15, vol. 14. 16 p. 7 pls. Paper.

Describing here the contributions made by Solon to the ceramic world, Hobson places particular emphasis on the former's use of the "pate sur pate" technique. All of the halftone plates illustrate the use of this technique on several vases and plaques.

Hodgson, Jeannette R. COLLECTING OLD ENGLISH LUSTRE. Portland, Maine: Southworth-Anthoensen Press, 1937. xv, 44 p. 14 pls.

Writing to stimulate people's interest in collecting, the author narrates some of her own collecting experiences and informally explains the history of English luster pottery, including the types produced and processes of manufacture. Several examples from the author's collection are illustrated in each plate.

Holgate, David. NEW HALL AND ITS IMITATORS. Faber Monographs on Pottery and Porcelain. London: Faber and Faber, 1971. xvi, 112 p. 265 illus. in 102 pls. (8 color). Bibliography, pp. 109-10. Index, pp. 111-12.

Holgate describes the New Hall factory as the last maker of hard-paste porcelain in England before it, too, was forced to turn to the more economic bone-china body. Included are brief histories of the factory and its wares, and more detailed information on decoration and marks. The second part of this volume examines New Hall imitators, some of whom, like Coalport and Minton, simply adapted New Hall designs while others, unidentified, copied the New Hall pieces anonymously.

Honey, W[illiam].B[owyer]. ENGLISH POTTERY AND PORCELAIN. 6th ed. Rev. by R.J. Charleston. The Library of English Art. London: A. & C. Black, 1969. xvi, 287 p. 24 pls. Bibliography, pp. 271-77. Index, pp. 278-87.

Intending his study to be a fairly concise history of English ceramic art, Honey has divided this volume into three parts. The first deals with English earthenware and stoneware to the end of the nineteenth century, the second with English porcelain of the eighteenth century, and the third with nineteenth-century pottery and porcelain in general. The author also refers the reader to his OLD ENGLISH PORCELAIN for a more detailed study of that subject.

_____. WEDGWOOD WARE. Faber Monographs on Pottery and Porcelain. London: Faber and Faber, 1948. 35 p. 100 pls. (4 color). Bibliography, p. 31. Index, pp. 33-35.

In this brief, illustrated history, the author attempts a more balanced view of Josiah Wedgwood's contribution, including examples of his work using colored glazes and the important cream-colored ware, as well as the more commonly known decorative stonewares.

Howard, David Sanctuary. CHINESE ARMORIAL PORCELAIN. London: Faber and Faber, 1974; distributed by Elinor Gordon, Box 211, Villanova, Pa. 19085. xiii, 1,034 p. 25 color pls., 41 halftones, numerous armorial illus. 14 appendixes, pp. 872-935. Addendum, pp. 938-1002. Bibliography, pp. 1003-5. 2 indexes, pp. 1007-34.

This enormous volume is basically structured like a dictionary of porcelain services produced for the British market from 1690 to 1820. Some 2,900 services are recorded and examples from most of them are illustrated. Almost every possible facet of these wares is treated here, including the manufacture of porcelain, the East India Company, the history of Canton ware, heraldry, and coats of arms. It would not seem an exaggeration to describe this text as the definitive study of these wares.

Hughes, G. Bernard. VICTORIAN POTTERY AND PORCELAIN. London: Country Life, 1959. 184 p. 90 illus. in 33 pls. Bibliography, p. 173. Index, pp. 174-84.

Designing his book for the collector and covering everything from bone china of the 1800s to art pottery of the 1890s, Hughes includes chapters on virtually every type of nineteenth-century English ware. Basalts, terra-cottas, ironstone, slipware, and many porcelain factories are each treated at least briefly.

Hurlbutt, Frank. BRISTOL PORCELAIN. London and Boston: Medici Society, 1928. xx, 164 p. 102 illus. in 72 pls. (8 color). Bibliography, p. 160. Index, pp. 161-64.

Specifically written for the collector, Hurlbutt's text covers the history of the Bristol factories through the eighteenth century and includes chapters on the technical composition of the wares as well as the types and styles of work produced. There is also a chapter on marks and numbers.

Hyde, J.A. [Lloyd]. ORIENTAL LOWESTOFT.

See: Chapter 7: "Ceramics of the United States, Canada, and Mexico" (p. 108).

Jewitt, Llewellynn. THE HISTORY OF THE CERAMIC ART IN GREAT BRITAIN. 2 vols. New York: Scribner, Welford, & Armstrong, 1878. Vol. 1: xiii, 531 p. 888 figs. Vol. 2: viii, 569 p. 786 figs. 3 indexes, pp. 539-69.

This old but quite extensive work covers the entire history of pottery in the British Isles from prehistoric times on. Hundreds of potters and factories are identified as are numerous styles and marks in engraved illustrations. A list of all ceramic patents is included at the end of the text.

_____. LIFE OF JOSIAH WEDGWOOD. London: Virtue Brothers & Co., 1865. xx, 435 p. 145 figs. Index, pp. 423-35.

Beginning with several chapters of material on the early potteries of Staffordshire, this biography describes Wedgwood's discoveries, the growth of his potteries, his difficulties, family life, etc. Jewitt also includes two chapters describing the lives of Wedgewood's ancestors and descendants.

John, W[illiam].D[avid]. WILLIAM BILLINGSLEY (1758-1828). Newport, Wales: Ceramic Book Co., 1968. xvi, 97 p. 180 illus. in 92 pls. (21 color). Bibliography, p. xvii. Index, pp. 129-32.

Large in format and handsomely illustrated with numerous color plates, this biography recounts the life of Billingsley from his years as a young flower painter at the Derby porcelain factory through his attempts to create a hard-paste body comparable to Sevres, and his final triumph as the maker of Nantgarw porcelain. His work as a painter and gilder at various factories is also included and his painting on Swansea porcelain is treated in the appendix.

John, W[illiam].D[avid]., and Baker, Warren. OLD ENGLISH LUSTRE POTTERY. Newport, Wales: R.H. Johns, 1951. xx, 132 p. 95 pls. 545 illus. (44 color). Bibliography, p. xvii. Index, pp. 129-32.

This large, well-illustrated volume explains the origins and techniques of lustered pottery and covers the various types and styles of ware as well as the factories which produced them. Included are Spode, Wedgwood, Staffordshire, Leeds, and Swansea. The work of William DeMorgan is also treated and two chapters are devoted to American historical lusterware.

Kelly, Alison. DECORATIVE WEDGWOOD IN ARCHITECTURE AND FURNITURE. London: Country Life, 1965. 147 p. 79 illus. (4 color). Glossary, pp. 135-36. Bibliography, pp. 137-39. Index, pp. 141-47.

Characterizing Josiah Wedgwood as an excellent businessman as well as a great artist, Kelly examines his work in terms of the crosscurrents of fashion in the late eighteenth century and Wedgwood's response to them. Concentrating on the jasper and basalt wares, the author illustrates chimneypieces, bridges, fonts, tiles, furniture, and various decorative pieces that use Wedgwood plaques as part of their ornamentation.

King, William. CHELSEA PORCELAIN. London: Benn Brothers, 1922. xv, 139 p. 71 pls. (7 color). Appendix, pp. 69-130. Bibliography, p. 131. Index, pp. 133-35.

King arranges his descriptions of the wares produced during each of the factory's periods according to their changing marks. The formulas of the different porcelain bodies used are described and the sale catalog of the factory's complete production during 1755 is also included. There are numerous illustrations of porcelain figures, service pieces, and marks.

______. ENGLISH PORCELAIN FIGURES OF THE EIGHTEENTH CENTURY SELECTED AND DESCRIBED. London: Medici Society, 1925. 15 p. 79 pls. (7 color). Bibliography, p. 15.

Beginning with a brief historical introduction, this small volume contains typical examples of eighteenth-century human figures in porcelain.

Laidacker, Sam, ed. THE STANDARD CATALOGUE OF ANGLO-AMERICAN CHINA: FROM 1810 TO 1850.

See: Chapter 7: "Ceramics of the United States, Canada, and Mexico" (p. 109).

Lancaster, H. Boswell. LIVERPOOL AND HER POTTERS. Liverpool: W.B. Jones & Co., 1936. 95 p. 30 illus. in 23 pls. References, p. 95. Index, pp. i-iii.

Working from old records and extracts, the author attempts to collate a brief history of the development of the pottery industry in the Liverpool area from the beginning of the eighteenth century. An extensive number of potteries and individuals are identified and several types of ware are examined.

Lane, Arthur. ENGLISH PORCELAIN FIGURES OF THE EIGHTEENTH CENTURY. Faber Monographs on Pottery and Porcelain. London: Faber and Faber, 1961. xi, 148 p. 136 illus. in 100 pls. (4 color). 2 appendixes, pp. 133-39. Bibliography, pp. 140-41. Index, pp. 143-48.

Intended as one of four monographs on German, English, French, and Italian porcelain figures, Lane's text describes the subjects, sources, relation to sculpture, and production techniques of these figures. Several chapters describe the figures produced by individual artists at a number of English porcelain factories.

Latham, Bryan. VICTORIAN STAFFORDSHIRE PORTRAIT FIGURES FOR THE SMALL COLLECTOR. London: Alec Tiranti, 1953. 41 p. 71 halftones in 56 pls.

This collector's volume surveys a variety of figures portraying famous individuals and events from the reign of Victoria. Grouping them according to subject matter, Latham includes royal figures, foreign royalty, Crimean War participants, the Indian Mutiny figures, celebrities, politicians, sportsmen, and even murderers. A note on each illustration describes the figure and identifies the individual or scene depicted.

Leeds, England. City Art Gallery. HANDBOOK OF THE LEEDS POTTERY AND CATALOGUE OF THE EXHIBITED MATERIAL. Leeds, 1951. 48 p. 19 halftones. Paper.

Written by Donald Towner of the English Ceramic Circle, this exhibition catalog covers the history of the Leeds pottery from its origins before 1771 to its period of decline in the early nineteenth century. Information on pattern and drawing books, marks, and glazes is included and each of the 131 pieces in the exhibit is identified and described.

Lewis, Griselda. AN INTRODUCTION TO ENGLISH POTTERY. London: Art & Technics, 1950. 40 p. 55 pls., 16 figs.

Although each type of pottery is treated here quite briefly, Lewis does touch upon all of the basic styles of English earthenware from the earliest Neolithic pottery pieces to the transfer-printed wares of the eighteenth and nineteenth centuries. The origin and manufacture of each type is explained and illustrated.

London. Royal Academy of Arts. 200 YEARS OF SPODE. London: 1970. 92 p. Illus. (5 color). Paper.

This is the catalog of an exhibition celebrating Spode china's bicentennial. Included are a short history, a photographic essay on the production process, and illustrations of a number of the exhibition pieces.

McCauley, Robert H. LIVERPOOL TRANSFER DESIGNS ON ANGLO-AMERICAN POTTERY.

See: Chapter 7: "Ceramics of the United States, Canada, and Mexico" (p. 109).

MacDonald, J.D. OLD ENGLISH CHINA AND POTTERY IN AUSTRALASIA. Sydney, Aust.: Shepherd Press, [1945]. 164 p. 133 halftones. Glossary, pp. 163-64. Index.

The author's intent is not to present a true history of English ceramic wares but to illustrate and describe a number of excellent examples of English porcelain produced between 1740 and 1820 and presently in Australian collections. While the emphasis here is on porcelain, the first chapter is devoted to earthenwares such as redware, salt glazes, and Toby jugs. The products of some fifteen English porcelain factories are identified and illustrated.

Macht, Carol. CLASSICAL WEDGWOOD DESIGNS. New York: M. Barrows & Co., 1957. xvi, 144 p. 64 illus. Bibliography, pp. 132-38. Index, pp. 139-44.

After briefly outlining Wedgwood's life and the main characteristics of the classical revival, Macht discusses Wedgwood's patrons, books, jewels, and other design sources. The exact sources for a number of particular designs are treated individually and are compared closely with Wedgwood's copies.

Mackenna, F. Severne. CHELSEA PORCELAIN: THE TRIANGLE AND RAISED ANCHOR WARES. Leigh-on-Sea, Eng.: F. Lewis, Publishers, 1948. xvi, 90 p. 113 illus. in 56 pls. Bibliography, pp. 75-77. Index, pp. 87-90.

In addition to describing the various wares produced at the Chelsea factory, including domestic objects, figures, fable-decorated wares, and the "girl in a swing" class, Mackenna briefly relates the history of the factory. The names of artists and workmen are given as well as fakes, marks, and the composition of wares. While numerous pieces are illustrated, all of the photographs have a uniformly unpleasant grayness, which spoils their clarity.

______. EIGHTEENTH CENTURY ENGLISH PORCELAIN: NOTES ON VARIOUS ASPECTS OF COLLECTING. Leigh-on-Sea, Eng.: F. Lewis, Publishers, 1970. xviii, 69 p. 102 illus. in 47 pls. Index, pp. 67-69.

This volume is a collection of previously published material by Mackenna touching on various aspects of porcelain history and giving general advice for the collector. Some thirteen topics are treated and there are numerous illustrations.

Mackenzie, Compton. THE HOUSE OF COALPORT, 1750-1950. London: Collins, 1951. 128 p. 33 illus. (15 color). Appendix, pp. 122-26. Marks, pp. 127-28.

This small volume is a commemorative history written for the Coalport factory's bicentennial celebration. Although it does record the moves of the factory and several styles of its wares, this very informally written book is probably of limited value except to a student of local history.

Mankowitz, Wolf. WEDGWOOD. London: Spring Books, 1966. xx, 284 p. 124 illus. (8 color). Glossary, pp. 203-9. Bibliography, p. 275. Index, pp. 279-84.

Concentrating on its experimental history and distinguishing qualities, this study examines Wedgwood pottery from its early colored glazed ware to the work of Emile Lessore in the 1860s. Also included is a transcript of the 1779 Wedgwood and Bentley catalog plus material added to it in 1787.

*Mankowitz, Wolf, and Haggar, Reginald G. THE CONCISE ENCYCLOPEDIA OF ENGLISH POTTERY AND PORCELAIN. New York: Frederick A. Praeger, 1957. 312 p. 184 pls. (24 color). 6 appendixes, pp. 250-99. Bibliography, pp. 300-311. (Chamberlin, no. 1739).

Clearly, alphabetically organized for quick reference by the collector, this book includes a good deal of information in short summary form concerning factories, manufacturers, artists, types of pottery and porcelain, and their styles and decoration.

Mayer, Joseph. HISTORY OF THE ART OF POTTERY IN LIVERPOOL. MEMOIRS OF THE HISTORIC SOCIETY OF LANCASHIRE AND CHESHIRE, vol. 7. Liverpool: T. Brakell, 1855. 37 p. Figs.

This early and informal history of the potteries at Liverpool briefly describes the origins of the various factories and the types of wares produced there during the seventeenth and eighteenth centuries.

Meager, Kildare S. SWANSEA AND NANTGARW POTTERIES. The Glynn Vivian Art Gallery, Swansea. Swansea, Wales: Swansea Corp., 1949. 51 p. 36 halftones in 18 pls. 2 appendixes, pp. 19-25. Catalog, pp. 26-51.

In addition to Meager's brief history of the Swansea and Nantgarw potteries, this publication contains a catalog of some 350 pieces of Welsh pottery and porcelain in the Glynn Vivian Art Gallery, Swansea. Appendix A consists of several recipes for porcelain bodies and glazes used by the Cambrian Pottery, Swansea.

Meteyard, Eliza. THE LIFE OF JOSIAH WEDGWOOD. 2 vols. London: Hurst & Blackett, Publishers, 1865. Vol. 1: xxiv, 504 p. 96 figs. Vol. 2: xxiv, 643 p. 154 figs. Index, pp. 621-43.

Covering practically every facet of Wedgwood's life and work, this detailed biography also includes a fairly extensive history of pottery in England from the earliest times to the work of Wedgwood's immediate predecessors at Staffordshire.

Mew, Egan. OLD BOW CHINA. Masterpieces of Handicraft. London: T.C. and E.C. Jack; New York: Dodd, Mead & Co., [1909]. 111 p. 16 pls. (8 color).

This old, collector's volume informally explores the history and products of the Bow factory, founded at Stratford-by-Bow around 1740. Most of the illustrations are of figures although some tablewares are also included.

Miller, J. Jefferson. ENGLISH YELLOW-GLAZED EARTHENWARE. Washington, D.C.: Smithsonian Institution Press, 1974. xvii, 125 p. 138 illus. (64 color). Bibliography, pp. 115-17. Appendix, pp. 119-20. Index, pp. 121-25.

Based on the Leon collection now in the possession of the Smithsonian, this well-illustrated study examines the styles, forms, and decoration of these wares from their origin at the end of the eighteenth century. A brief chapter is also devoted to examples of these wares decorated with American views.

Moore, N. Hudson. THE OLD CHINA BOOK. New York: Tudor Publishing Co., 1903. 300 p. 150 halftones in 74 pls. List of views, pp. 255-83. Bibliography, p. 284. Index, pp. 285-300.

Intended as a guide for the collector, Moore's text touches on the history and styles of English pottery and porcelain and treats the Liverpool printed wares, porcelains, basalts, lusters, and whiteware. Additional chapters cover Wedgwood pieces and the work of a number of the major Staffordshire potteries.

Mountford, Arnold R. THE ILLUSTRATED GUIDE TO STAFFORDSHIRE SALT-GLAZED STONEWARE. Illustrated Guides to Pottery and Porcelain, Geoffrey A. Godden, general ed. New York: Praeger Publishers, 1971. xxi, 88 p. 128 pls. 251 illus. (7 color). 2 appendixes, pp. 69-77. Bibliography, pp. 78-79. Index, pp. 81-88.

Incorporating a tremendous number of illustrations, this history deals with the origin and varying styles of Staffordshire salt-glazed ware. The guide also contains descriptions and illustrations of pieces recovered from excavations at several eighteenth-century pottery sites.

Nicholls, Robert. TEN GENERATIONS OF A POTTING FAMILY. London: Percy Lund, Humphries & Co., [n.d.]. xxvi, 135 p. 53 pls., geneaology. Glossary, pp. xvii-xviii. Marks, pp. xix-xxi. 3 indexes, pp. 121-35.

This history of the Adams family is based on two earlier volumes, WILLIAM ADAMS: AN OLD ENGLISH POTTER, by William Turner, and A HISTORY OF THE ADAMS FAMILY OF NORTH STAFFORDSHIRE. Using these two books and additional material, the author covers the history of the Adams potteries from the early records of the Adams family in the fifteenth century through the products of the various Adams master potters in the eighteenth and nineteenth centuries.

Oliver, Anthony. THE VICTORIAN STAFFORDSHIRE FIGURE: A GUIDE FOR COLLECTORS. New York: St. Martin's Press, 1971. xiii, 179 p. 8 color pls., 250 halftones. Bibliography, p. 173. Index, pp. 175-79.

In addition to lengthy ones on identification, forgeries, and general collecting, Oliver devotes several chapters to background information on these nineteenth-century pieces, discussing the people who made them and how and of what they were made. Hundreds of the figures and a number of the contemporary engravings and paintings that probably had inspired them are illustrated.

Oxford University Delegacy for Extra-Mural Studies. "Lectures on Pottery." [Oxford]: 1958. 58 p. Paper. Mimeographed.

These four lectures were delivered during 1958 and arranged in consultation with the North Staffs, Extra-Mural Joint Advisory Committee, and the City Museum, Stoke-on-Trent. The titles and authors of the lectures are as follows: "Tradition and Taste in English Ceramics" by G.J.V. Bemrose; "English Cream-Coloured Earthenware" by John Lowe; "Staffordshire Pottery Figures--Past and Present" by Reginald Haggar; and "English Salt-Glazed Stoneware" by Arthur Lane. While slides were presented with the lectures, they are not listed or printed here.

Pugh, P.D. Gordon. STAFFORDSHIRE PORTRAIT FIGURES AND ALLIED SUBJECTS OF THE VICTORIAN ERA. New York: Praeger Publishers, 1971. xi, 657 p. Illus. (part color). Bibliography, pp. 645-47. Index, pp. 651-57.

This profusely illustrated, comprehensive study includes a general history of these nineteenth-century figures and examines their sources, methods of manufacture, and the numerous factories that produced them. An extensive catalog illustrates what must be at least a thousand examples and each character is identified in a descriptive paragraph. For all practical purposes, this is the definitive text on these figures for the collector.

Rackham, Bernard. MEDIEVAL ENGLISH POTTERY. Rev. ed. Faber Monographs on Pottery and Porcelain. London: Faber and Faber, 1972. xvi, 40 p. 107 illus. in 104 pls. (8 color). Bibliography, pp. 33-35. Index, pp. 37-39.

Rackham's original edition of this book, published in 1948, was the first volume ever devoted to this topic. Rackham describes these earthenwares as object lessons teaching the essential ceramic virtues and compares their forms favorably to those of early Chinese wares. He further argues that their forms are unmistakably English and that all later English styles have been indebted to them.

Rackham, Bernard, and Read, Herbert. ENGLISH POTTERY: ITS DEVELOPMENT FROM EARLY TIMES TO THE END OF THE EIGHTEENTH CENTURY. New York: Charles Scribner's Sons, 1924. xxiv, 143 p. 115 pls. (12 color). Bibliography, p. xxiv. Index, pp. 139-43.

This critical history examines the various influences on the development of English pottery. Chapters are devoted to foreign strains, the rise of industrialism, and the influence of porcelain. An appendix on the Wrotham potters by J.W.L. Glaisher is included.

Ray, Anthony. ENGLISH DELFTWARE TILES. London: Faber and Faber, 1973. 287 p. 88 pls. (8 color), 35 figs. Bibliography, pp. 281-82. Index, pp. 283-87.

This text deals with the history, manufacture, decoration, and use of delftware tiles in England with particular reference to contemporary English and Dutch documents. Intended as a reference book, the volume shows some eight hundred tile designs arranged according to subject. The products of such centers as London, Bristol, and Liverpool are differentiated whenever possible and a separate section is devoted to the work of John Sadler and his use of transfer printing on tiles.

Rhead, G. Wooliscroft, and Rhead, Frederick Alfred. STAFFORDSHIRE POTS AND POTTERS. London: Hutchinson & Co., 1906. 384 p. 4 color pls., 116 halftones, 90 figs. Index, pp. 373-84.

Intended as an informal history of the potter's art in the Staffordshire area, this text begins with the pre-Roman potters of Britain and continues up into the nineteenth century. The authors attempt to acquaint the reader "with the personality of the more considerable potters and the conditions under which they worked."

Rice, Dennis G. THE ILLUSTRATED GUIDE TO ROCKINGHAM POTTERY AND PORCELAIN. Illustrated Guides to Pottery and Porcelain, Geoffrey A. Godden, general ed. New York and Washington, D.C.: Praeger Publishers, 1971. xx, 194 p. 179 illus. in 112 pls. (8 color). 7 appendixes, pp. 115-38. Bibliography, pp. 186-87. Index, pp. 189-94.

Beginning with a brief history of Rockingham pottery and porcelain manufacture, Rice devotes separate chapters to the pottery, useful porcelain, ornamental porcelain, figures, artists, and marks of Rockingham. The appendixes cover such items as the Rhinoceros Vases, pseudo-Rockingham figures, the London House, and the Wentworth correspondence and include a list of the figures.

Sandon, Henry. ROYAL WORCESTER PORCELAIN FROM 1862 TO THE PRESENT DAY. London: Barrie & Jenkins, 1973. xxix, 265 p. 238 illus. in 130 pls. (18 color). Glossary, pp. xxii-xxix. Index, pp. 257-65.

This is more accurately a company history than a history of Worcester porcelain. Detailed information on workers, changes at the factory, and new products is included for each decade of the company's existence starting in the 1860s. Hundreds of workers are identified and thousands of numbers referring to painters, gilders, patterns, and shapes are explained.

Sempill, Cecilia. ENGLISH POTTERY AND PORCELAIN. Britain in Pictures; W.J. Turner, general ed. London: Collins, 1944. 48 p. 37 illus. (8 color), marks.

This brief popular history examines the individual and unique qualities of various English pottery and porcelain pieces from medieval times to the twentieth century.

Shaw, Simeon. HISTORY OF THE STAFFORDSHIRE POTTERIES. Hanley, Eng.: 1829. Reprint. New York: Praeger Publishers, 1970. 244 p. 2 indexes, pp. 239-44.

This history recounts the development of the Staffordshire pottery area from the seventeenth century on. The manufacturers and such products of the eighteenth century as porcelain, red porcelain, Queen's ware, blue-printed pottery, and luster pottery are also discussed.

Shinn, Charles, and Shinn, Dorrie. THE ILLUSTRATED GUIDE TO VICTORIAN PARIAN WARE. Illustrated Guides to Pottery and Porcelain. London: Barrie & Jenkins, 1971. xv, 125 p. 119 illus. in 98 pls. (2 color). Index, pp. 123-25.

Shinn's guide describes the origins and development of nineteenth-century British Parian ware, the role of the art unions, and the principal manufacturers and their methods. Several chapters offer advice on collecting as well as extensive lists of sculptors, makers, and their marks.

Smiles, Samuel. JOSIAH WEDGWOOD, F.R.S.: HIS PERSONAL HISTORY. New York: Harper & Brothers, Publishers, 1895. 330 p. Index, pp. 327-30.

Though described as a personal history, this book discusses the developments in Wedgwood's professional as well as personal life. While this biography is extensive and appears to be reasonably complete, it does not have either the detail or scope of Meteyard's larger study (see p. 79).

Smith, Alan. THE ILLUSTRATED GUIDE TO LIVERPOOL HERCULANEUM POTTERY, 1796-1840. The Illustrated Guides to Pottery and Porcelain. New York: Praeger Publishers, 1970. xvi, 142 p. 191 illus. in 104 pls. (8 color), 8 figs., 4 maps. 3 appendixes, pp. 108-31. Bibliography, pp. 132-33. Index, pp. 135-42.

Attempting to clarify the history of the Liverpool wares produced between 1796 and 1840 at the Herculaneum factory, the author treats the factory's various stages of growth and eventual decline, types of wares, and methods of production. A separate chapter deals with the overseas trade and the appendixes contain information on marks and factory records.

Solon, M.L. [Louis Marc]. THE ART OF THE OLD ENGLISH POTTER. 2d ed., rev. London: Bemrose & Sons, 1885. xxiv, 269 p. 55 drwgs. Index, pp. 267-69.

This antique text is perhaps most aptly described as a gentlemanly history of English pottery from prehistoric times to the eighteenth century. The illustrations are from hand-etched plates executed by the author.

______. A BRIEF HISTORY OF OLD ENGLISH PORCELAIN AND ITS MANUFACTURIES. London: Bemrose & Sons, 1903. xvi, 256 p. 50 pls. (20 color). Bibliography, pp. 243-47. 2 indexes, pp. 249-56.

Beginning with the early years of porcelain manufacture in England, Solon describes the history and products of each of the English porcelain factories, including some of the lesser-known and disputed factories, such as Pinxton & Torksey, Church Gresley, Liverpool, and Davenport. He discusses the marks of several factories and makes a critical appraisal of their products.

Spelman, W.W.R. LOWESTOFT CHINA. London: Jarrold & Sons, 1905. 78 p. 97 pls. (part color).

Spelman attempts to clear up the mystery of the china produced at Lowestoft by reporting the results of his discoveries at the factory site during excavations between 1902 and 1904. He is of the opinion that no hard-paste porcelain was made at Lowestoft and that "Oriental Lowestoft" is genuinely oriental.

THE STORY OF WEDGWOOD: A LIVING TRADITION. Rev. ed. Barlaston, Stoke-on-Trent, Eng.: Josiah Wedgwood & Sons, 1946. 51 p. Illus. Bibliography, p. 51. Paper.

This company pamphlet provides a short history of the Wedgwood family and its pottery factory.

Tapp, William H., M.C. JEFFERYES HAMETT O'NEALE, RED ANCHOR FABLE PAINTER, AND SOME CONTEMPORARIES. London: University of London Press, 1938. 66 p. 35 pls. (1 color). Index, pp. 61-66.

Tapp's study examines the work of J.H. O'Neale, the principal painter of scenes from fable, myth, and classic for the Chelsea factory during the "red anchor" period (1751-58). Tapp also includes information on Fidello and Joseph Michel DuVivier, who imitated O'Neale's work.

Tiffany & Co. THE BARBERINI VASE: DISCOVERY, COMPOSITION AND DESCRIPTION OF PORTLAND'S MYSTIC URN. New York: 1904. 53 p. 6 halftones.

This small volume on the Portland Vase includes several illustrations of its four sides and bottom as well as a description of Josiah Wedgwood's work to reproduce the vase in clay.

Tilley, Frank. TEAPOTS & TEA. Newport, Wales: Ceramic Book Co., 1957. 130 p. 205 illus. in 67 pls. (10 color). Index, pp. 133-35.

This coffee table volume covers the origins and development of pottery and porcelain teapots in England from their beginnings to the end of the eighteenth century. A major part of the text is devoted to descriptions of each illustration and an additional part explains porcelain analyses. Several aspects of the history of tea in England are also touched upon in a short chapter.

Towner, Donald C. ENGLISH CREAM-COLOURED EARTHENWARE. Faber Monographs on Pottery and Porcelain. London: Faber and Faber, 1957. xv, 107 p. 155 illus. in 100 pls. (4 color). 3 appendixes, pp. 64-100. Bibliography, p. 101. Index, pp. 103-7.

Beginning with a chapter on the pioneers of this ware, Towner writes separate chapters on the Leeds, Wedgwood, Swinton, and later Yorkshire and Staffordshire potteries. A brief chapter on continental creamware is also included and the appendixes contain numerous figure illustrations of handle and spout designs, patterns, and marks.

Turner, H.A.B. A COLLECTOR'S GUIDE TO STAFFORDSHIRE POTTERY FIGURES. London: MacGibbon & Kee, 1971. xv, 294 p. 8 color pls., 88 halftones. Bibliography, pp. 276-79. Index, pp. 281-94.

Covering the period from 1740 to 1900, this text describes a number of the individuals, schools, and styles related to these pottery figures. Separate chapters are devoted to Victoriana, animals, identifiable factories, manufacturing techniques, Toby jugs, lists of potters, and museums with Staffordshire holdings.

Watkins, C. Malcolm. NORTH DEVON POTTERY AND ITS EXPORT TO AMERICA IN THE SEVENTEENTH CENTURY. United States National Museum Bulletin, no. 225. Washington, D.C.: Smithsonian Institution, 1960. 44 p. 1 color pl., 35 halftones and figs. Paper.

Using evidence from excavations of sites such as Plymouth, Massachusetts, and Jamestown, Virginia, the author examines the trade in pottery between England and her American colonies.

Watney, Bernard. ENGLISH BLUE AND WHITE PORCELAIN OF THE EIGHTEENTH CENTURY. Faber Monographs on Pottery and Porcelain. Faber and Faber, 1963. xiii, 137 p. 322 illus. in 100 pls. (4 color). Bibliography, pp. 130-32. Index, pp. 133-37.

Explaining that blue and white ware was generally easier and cheaper to produce, and therefore more commonly made for daily use than the difficult colored wares, Watney describes the various styles and production techniques of each English factory that produced it. Almost every English porcelain factory is cited here.

_____. LONGTON HALL PORCELAIN. Faber Monographs on Pottery and Porcelain. London: Faber and Faber, 1957. xvi, 72 p. 155 illus. in 84 pls. (4 color). Marks, p. 49. Bibliography, pp. 50-51. 2 appendixes, pp. 52-69. Index, pp. 70-72.

Watney describes the brief history of one of the first English porcelain factories, which began in 1750 and ended in financial failure in 1760. The entire history of this Staffordshire factory was only rediscovered in recent times by collectors. Chapters on the factory's early, middle, and late wares are included.

Wedgwood, Josiah C. STAFFORDSHIRE POTTERY AND ITS HISTORY. New York: McBride, Nast & Co., 1913. 129 p. 26 illus., maps, charts. Appendixes, pp. 213-21. Index, pp. 223-29.

Written by the great-great-grandson of Josiah Wedgwood (1730-95), and seemingly intended as a local history, this volume includes a great many area names. As a result, it serves as a family chronicle, an informal history of the region, and a document of the growth of the pottery industry.

Whiter, Leonard. SPODE: A HISTORY OF THE FAMILY, FACTORY AND WARES FROM 1733 TO 1833. London: Barrie & Jenkins, 1970. xiii, 246 p. 89 pls. (9 color), figs. 4 appendixes, pp. 200-227. Bibliography, pp. 236-39. 2 indexes, pp. 240-46.

Handsomely illustrated with a large selection of plates and hundreds of shapes and patterns from the factory records, this history covers the work of the two Josiah Spodes and the Copeland family. The growth of the business, the decorative work of Henry Daniel, and the types of wares produced are all treated. Information on pricing, factory marks, and the members of the Spode family is included in the appendixes.

Williams, Sydney B. ANTIQUE BLUE AND WHITE SPODE. London: B.T. Batsford, 1943. 242 p. 122 illus. (bluetone). Index, pp. 240-42.

Dealing principally with the wares produced in the first quarter of the nineteenth century by the younger Josiah Spode, this collector's volume includes information on the history of the period, dates of the wares, contemporary engravings used as models, and various influences on the different patterns and designs used by the company on their blue-and-white ware.

Williams-Wood, Cyril. STAFFORDSHIRE POT LIDS AND THEIR POTTERS. London: Faber and Faber, 1972. 173 p. 81 pls. (9 color), figs. 3 appendixes, pp. 107-61. Glossary, pp. 162-64. Bibliography, p. 165. Index, pp. 167-73.

In addition to describing the origins of these early commercially produced jar covers and the techniques of transfer printing on pottery, the author details the characteristics, subject matter, and types of lids produced by some ten pottery companies starting in the early nineteenth century. Several lists of all the recorded polychrome and monochrome pictures found on lids are included in the appendixes.

*Wills, Geoffrey. ENGLISH POTTERY AND PORCELAIN. Garden City, N.Y.: Doubleday & Co., 1969. 291 p. Illus. Bibliography. Index, pp. i-vii.

Profusely illustrated in black and white and color, Wills's very readable history covers the pottery and porcelain of England from the early red earthenwares of the sixteenth and seventeenth centuries to the porcelains of the late nineteenth century. Delftware, stoneware, creamware, Wedgwood, and the products of the various porcelain centers are discussed.

Woodhouse, Charles Platten. OLD ENGLISH TOBY JUGS AND THEIR MAKERS. London: Mountrose Press, 1949. 48 p. 12 halftones.

This informal history of a unique British pottery form recounts the origins and history of these pieces from their beginnings in the eighteenth century, and cites the work of several of its most famous makers. A number of examples are illustrated and the author gives advice to potential collectors.

EUROPEAN POTTERY AND PORCELAIN

Auscher, E[rnest].S[imon]. A HISTORY AND DESCRIPTION OF FRENCH PORCELAIN. Translated by William Burton. London: Cassell & Co., 1905. xiv, 200 p. 84 pls. (24 color). Marks and monograms, pp. 173-91. Bibliography, pp. 192-96. Index, pp. 197-200.

"Chef de Fabrication" at the Sevres factory from 1879-89, Auscher covers the entire history of porcelain manufacture in France from the invention of soft paste to the restoration of the Sevres factory by Alexandre Brongniant during the First Empire. Auscher's concluding chapters are devoted to modern developments, forgeries, and the markings used throughout the history of porcelain manufacture in France.

Avery, C. Louise. MASTERPIECES OF EUROPEAN PORCELAIN. New York: Metropolitan Museum of Art, 1949. 30 p. 33 halftones in 32 pls. Paper.

This exhibition catalog includes a five-page introduction by Avery and a brief description of each of the 382 pieces in the exhibit. Included are examples of Austrian, English, French, German, Italian, and Spanish porcelain.

Bakurjiev, Georgi. BULGARIAN CERAMICS. Sofia, Bulg.: Bulgarski Hudozhnik, 1955. 40 p. 160 pls. (16 color).

This brief, popular history covers the development of and various influences on Bulgarian pottery from prehistoric times to the present. A large selection of illustrations are included, but a number are not of the best quality.

Barber, Edwin Atlee. SALT GLAZED STONEWARE: GERMANY, FLANDERS, ENGLAND AND THE UNITED STATES. Art Primer Series. Pennsylvania Museum and School of Industrial Art. New York: Doubleday, Page & Co., 1907. 32 p. 48 illus. Index, pp. 31-32.

This volume serves as a very brief introduction to the origins and characteristics of salt-glazed stoneware.

_____. TIN ENAMELLED POTTERY: MAJOLICA, DELFT AND OTHER STANNIFEROUS FAIENCE. Art Primer Series. Pennsylvania Museum and School of Industrial Art. New York: Doubleday, Page & Co., 1907. 51 p. 57 illus. Pottery marks, pp. 37-44. Index, pp. 49-51.

This brief text contains historical background, an explanation of processes, and descriptions of characteristic examples for both collectors and students.

Barton, Kenneth James. SOME EXAMPLES OF MEDIEVAL GLAZED EARTHENWARE IN SWEDEN. N.p.: Kungl. Vitterhets Historie och Antikvitets Akademien, Antikvariskt Arkiv 33. Distributed by Almqvist & Wiksell, Stockholm, 1968. 48 p. Halftones, figs. Paper.

Confining himself to jugs and special types of wares (aquamaniles), Barton examines the less common lead-glazed earthenwares in several Swedish museum collections. Individual pieces are described and locations, sources, types, and dating are discussed.

Beurdeley, Michel. CHINESE TRADE PORCELAIN.

See: Chapter 4: "Eastern Ceramics" (p. 27).

Bodelsen, Merete. GAUGUIN'S CERAMICS: A STUDY IN THE DEVELOPMENT OF HIS ART. London: Faber and Faber, 1964. 239 p. 162 illus. (part color).

Grouping some sixty ceramic pieces according to subject and style, the author tries to connect these known ceramics by Gauguin with his painting, drawing, and sculpture. Gauguin's work is treated chronologically and many of his ceramic pieces are illustrated in color. Several appendixes, lists of works, notes, and catalogs complete the text.

*Charles, Rollo. CONTINENTAL PORCELAIN OF THE EIGHTEENTH CENTURY. London: Ernest Benn, 1964. 198 p. 148 illus. in 76 pls. (12 color), map. Bibliography, p. 187. Index, pp. 189-98.

Excluding the wares produced in England during the eighteenth century, this history briefly surveys the unique contributions of the different porcelain factories throughout Europe. In addition to introductory chapters on porcelain and processes of production, separate chapters are devoted to individually famous porcelain factories and to other less famous factories in Switzerland, Holland, Scandinavia, and Russia.

Clouzot, Henri. ANDRE METTHEY: DECORATEUR ET CERAMISTE. Paris: Librairie des Arts Decoratifs, n.d. Illus. Portfolio.

This collection of forty-five halftone and color plates documents the work of the French ceramist Metthey (1871-1920).

De-Mauri, L. [Ernesto Sarasino]. VINOVO AND ITS PORCELAIN. Monographs of Decorative Arts, no. 4. Milan: Bottega de Poesia, 1925. 39 p. 34 pls. Bibliography, pp. 35-36.

This brief monograph includes a short history of this Piedmontese village and describes the work of the chemist Victor Amedeus Gioanetti. Vinovo's factory, different periods of manufacture and its porcelain's forms, decoration, special characteristics, and marks are examined while figures as well as functional and decorative pieces are illustrated.

M.H. DeYoung Memorial Museum. CONTINENTAL TABLE PORCELAINS OF THE EIGHTEENTH CENTURY. San Francisco: 1965. Unpaged. Halftones. Glossary, pp. 19-22. Bibliography, p. 22. Paper.

This exhibition catalog includes a brief history of the discovery and manufacture of soft- and hard-paste porcelain in the major European centers until the end of the eighteenth century. Each of the 284 items in the exhibition is identified and briefly described.

Doat, Taxile. GRAND FEU CERAMICS: A PRACTICAL TREATISE ON THE MAKING OF FINE PORCELAIN AND GRES. Translated by Samuel E. Robineau. Syracuse, N.Y.: Keramic Studio Publishing Co., 1905. Illus. (1 color), figs.

This text deals with the making and decorating of both high-fire porcelain and stoneware (gres) and includes technical information on clay bodies, glazes, and firing. A final section, by Professor Charles Binns, identifies the American clays which may be substituted for European clays in creating this ware.

Domanovszky, Gyorgy. HUNGARIAN POTTERY. Budapest: Corvina Press, 1968. 78 p. 48 pls. (16 color), map. Bibliography, p. 75.

This small, popular history of Hungarian peasant pottery includes information on its techniques and styles while individual chapters deal with regional pottery production in the Tisza region, the North Country, the Great Plain, Transdanubia, and Transylvania.

Frothingham, Alice Wilson. TILE PANELS OF SPAIN: 1500-1650. New York: Hispanic Society of America, 1969. 106 p. 104 illus. (6 color). Bibliography, pp. 93-95. Index, pp. 99-106.

Concentrating on the tile panels originating mainly at Seville and Talavera de la Reina, this volume explains the use of decoration and the Italian method of glaze painting during the Renaissance. Tile panels in the New World are also discussed.

Godden, Geoffrey A. ANTIQUE GLASS AND CHINA: A GUIDE FOR THE BEGINNING COLLECTOR. New York: A.S. Barnes & Co., 1966. 94 p. 47 halftones. Bibliography, p. 92. Index, pp. 93-94.

A general introduction to the collecting of various types of less expensive nineteenth-century English glass and china, this volume has additional chapters dealing with marks and registered patterns.

Haggar, Reginald G. THE CONCISE ENCYCLOPEDIA OF CONTINENTAL POTTERY AND PORCELAIN. London: Andre Deutsch, 1960. 533 p. 184 pls. (24 color). English bibliography, pp. 523-28. Foreign bibliography, pp. 529-33.

Designed as a companion volume to Mankowitz and Haggar's THE CONCISE ENCYCLOPEDIA OF ENGLISH POTTERY AND PORCELAIN (p. 79), this text is arranged alphabetically and identifies thousands of potters, factories, styles, and marks up to and including those of the twentieth century.

*Hillier, Bevis. POTTERY AND PORCELAIN 1700-1914: ENGLAND, EUROPE AND NORTH AMERICA. The Social History of the Decorative Arts, Hugh Honour, general ed. New York: Meredith Press, 1968. 386 p. 217 halftones, 16 color pls. Bibliography, pp. 361-68. Index, pp. 369-86.

This is a very readable, popular history of western ceramics from the end of the rococo to the art nouveau period. Its approach, however, is not encyclopedic. Rather, Hillier selects several major types of ceramic wares and treats them in some detail in their historical contexts.

Honey, W[illiam].B[owyer]. DRESDEN CHINA: AN INTRODUCTION TO THE STUDY OF MEISSEN PORCELAIN. London: A. & C. Black, 1934. 223 p. 61 halftones. Bibliography, pp. 174-76. Index, pp. 211-23. (Chamberlin, no. 1764).

This popular history of Meissen porcelain in the eighteenth century is designed for the use of the collector. Its emphasis is on identification and an understanding of the works in their historical context.

_____. EUROPEAN CERAMIC ART FROM THE END OF THE MIDDLE AGES TO ABOUT 1815: A DICTIONARY OF FACTORIES, ARTISTS, TECHNICAL TERMS AND GENERAL INFORMATION. London: Faber and Faber, 1952. 780 p. 4 color pls., figs., maps, marks. Bibliography, pp. 17-27. Index to marks, pp. 683-780.

Designed as a dictionary and an encyclopedia, this volume consists of brief entries arranged alphabetically for clear access. Marks are treated in the body of the text but also make up a separate index. This is a companion volume to Mr. Honey's ILLUSTRATED HISTORICAL SURVEY.

_____. EUROPEAN CERAMIC ART FROM THE END OF THE MIDDLE AGES TO ABOUT 1815: ILLUSTRATED HISTORICAL SURVEY. London: Faber and Faber, 1949. 85 p. 216 pls. (24 color). Index to plates, pp. 80-83. (Chamberlin, no. 1754.)

This well-illustrated volume offers an introductory survey of the various styles of European ceramics and is intended as a companion volume to Honey's DICTIONARY although the latter was published three years later.

_____. GERMAN PORCELAIN. Faber Monographs on Pottery and Porcelain, edited by W.B. Honey. London: Faber and Faber, 1947. 56 p. 100 illus. (4 color). Marks, pp. 37-47. Bibliography, pp. 49-51. Index, pp. 53-56.

Beginning with a description of the origins of European porcelain and its techniques of manufacture, Honey briefly discusses the products of some eighteen of the major factories, including Meissen, Vienna, Hausmalerei, Hochst, and Nymphenburg.

*Jonge, C[aroline].H[enriette]. de. DELFT CERAMICS. Translated by Marie-Christine Hellin. New York: Praeger Publishers, 1970. 168 p. 176 illus. (20 color). Delft potteries and marks, pp. 151-61. Bibliography, p. 168. 3 indexes, pp. 162-68.

This extensive, but not overly detailed, history of delft ceramics includes an explanation of techniques, the early transition from majolica to delft faience, oriental influences, and the various styles of production until its period of general decline in the eighteenth century.

_____. DUTCH TILES. Translated by P.S. Falla. New York: Praeger Publishers, 1971. 337 p. 400 illus. (154 color). Index, pp. 330-37.

This volume encompasses the development of ceramic tiles in The

Netherlands from the sixteenth to the nineteenth centuries and examines their various styles and methods and major centers of production.

Kiss, Akos. BAROQUE POTTERY IN HUNGARY. Budapest: Corvina Press, 1966. 49 p. 48 pls. (8 color). Bibliography, pp. 39–40.

This small history briefly examines the products of the two most outstanding manufacturers of faience in Hungary during the eighteenth century, Holics and Tata. A history of both factories and descriptions of influences on the products of each are included.

Knowles, W. Pitcairn. DUTCH POTTERY AND PORCELAIN. London: B.T. Batsford; New York: Charles Scribner's Sons, n.d. 122 p. 54 pls. 2 appendixes, pp. 71-111. Index, p. 122.

This brief, popular history explains the development of the various types of Dutch ceramics. One appendix lists the delft potteries and a second is comprised of the marks and monograms of the Guild of St. Luke.

Korf, Dingeman. DUTCH TILES. Translated by Marieke Clarke. New York: Universe Books, 1964. 136 p. Halftones, 360 figs. Bibliography, p. 135.

Although this small, popular history takes into account the development and manufacture of Dutch tiles, it is, for the most part, devoted to the enormous variety of patterns, scenes, and figures with which the tiles were decorated.

Krisztinkovich, Bela. HABAN POTTERY. Budapest: Corvina Press, 1962. 48 p. 55 illus. in 48 pls. (8 color). Bibliography, p. 42.

Examining the history of the majolica pottery produced during the sixteenth and seventeenth centuries in Hungary by the Habans, or Anabaptists, this small volume also touches upon the history of this group of people and the shapes, production, and methods of decoration of their pottery.

Lane, Arthur. FRENCH FAIENCE. Faber Monographs on Pottery and Porcelain. London: Faber and Faber, 1948. 49 p. 188 illus. in 100 pls. (4 color). Bibliography, p. 45. Index, pp. 47–49.

Lane surveys the development of French faience from its beginnings, inspired by Italian majolica, through its perfection as an art form, until it was driven from the market by the introduction of Wedgwood's hard, efficient earthenware in France. The wares produced in Nevers, Rouen, Moustiers, Marseilles, Strasbourg, Aprey, and Sceaux are all described.

_____. ITALIAN PORCELAIN. Faber Monographs on Pottery and Porcelain. London: Faber and Faber, 1954. xvi, 79 p. 195 illus. in 100 pls. (4 color). Bibliography, pp. 74-75. Index, pp. 77-79.

Covering the development of Italian porcelain from the late sixteenth to the eighteenth centuries, Lane's history discusses Medici porcelain (the first soft-paste porcelain produced in Europe), the Venetian factories, Vinovo, Doccia, and the Bourbon factories at Capodimonte, Buen Retiro, and Naples. A separate chapter is devoted to marks.

Litchfield, Frederick. POTTERY AND PORCELAIN: A GUIDE TO COLLECTORS. 5th ed. Revised by Frank Tilley. New York: M. Barrows & Co., 1951. 356 p. 96 pls. (32 color). Glossary, pp. 320-29. Bibliography, pp. 331-34. Index, pp. 335-56. (Chamberlin, no. 1755).

Concentrating almost exclusively on Europe, this volume focuses on the work of its various ceramic factories and their marks and monograms.

Moore, N. Hudson. DELFTWARE: DUTCH AND ENGLISH. Collector's Handbooks. New York: Frederick A. Stokes Co., 1908. 78 p. 57 illus. in 13 pls. Index, pp. 75-78.

This small, popular history briefly describes the development of delftware in both Holland and England and includes a twenty-four-page list of seventeenth- and eighteenth-century delft potters with reproductions of their marks.

Morley, Henry. THE LIFE OF BERNARD PALISSEY, OF SAINTES. 2 vols. London: Chapman & Hall, 1852. Vol. 1: 320 p. Vol. 2: 352 p. Index, pp. 349-52.

Palissey unsuccessfully attempted to develop a type of European porcelain during the sixteenth century. His work passed through several stages, including a superior type of peasant pottery decorated with applied reliefs and colored with glaze and enamels. He worked in Paris for about twenty-five years as a personal favorite of Catherine de Medici and her sons. This biography contains a good deal of information about Palissey's theories and work as a man of letters and a potter and includes selections from his writings. There are no illustrations.

Morley-Fletcher, Hugo. ANTIQUE PORCELAIN IN COLOR: MEISSEN. Garden City, N.Y.: Doubleday & Co., 1971. 119 p. Illus. (color).

Written by a director of Christie's and handsomely illustrated by pieces from private collections, this volume describes the discovery of porcelain in Europe and traces the history of the Meissen factory from 1710 to the nineteenth century.

Neurdenburg, Elizabeth. OLD DUTCH POTTERY AND TILES. Translated by Bernard Rackham. London: Benn Brothers, 1923. xv, 155 p. 112 illus. (9 color) in 69 pls. Appendix, pp. 145-46. Bibliography, pp. 147-48. Index, pp. 151-55. (Chamberlin, no. 1769).

Containing a large number of illustrations and covering the history of Dutch pottery from its red earthenware beginnings in the fourteenth century to its decline in the eighteenth century, this source includes chapters on Dutch majolica, tiles, tile-pictures, the origins of delft and polychrome delft, several famous potters, and the influence of true porcelain. An appendix provides a list of Delft potters in 1764.

Penkala, Maria. EUROPEAN PORCELAIN. Amsterdam: R.W. Haentjens Dekker, 1947. 314 p. Marks, pp. 177-302. Bibliography, pp. 303-8.

Containing brief summaries of each porcelain factory or town, this collector's handbook focuses on thousands of porcelain marks, which comprise at least one-half of the text.

Pica, Agnoldomenico. ITALIAN MAJOLICA TILE. Translated by James Pallas. Rome: Carlo Bestetti Edizioni D'Arte, 1969. 205 p. Illus. (part color). Bibliography, p. 79. Index, p. 205.

Published in Italian, French, German, and English, this volume contains an excellent selection of color and halftone illustrations. Part 1 consists of an informal text on the history of Italian tiles, while part 2, entitled "Italian Majolica Tile, Today," consists of large illustrations of contemporary tiles and tile panels. Except for its brief captions, part 2 contains no text.

Plenival de Guillebon, Regine de. PORCELAIN OF PARIS 1770-1850. Translated by Robin R. Charleston. New York: Walker & Co., 1972. 362 p. 243 illus. including 28 color pls. Sources, pp. 349-50. Bibliography, pp. 351-53. 2 indexes, pp. 355-61.

This large and handsomely illustrated volume includes in its scope all hard-paste porcelain made in Paris, although where exactly the pieces were made becomes more general in the nineteenth century. Styles, techniques, exhibitions, trademarks, and a substantial history of the factories are all incorporated.

Rackham, Bernard. ITALIAN MAJOLICA. 2d ed. Faber Monographs on Pottery and Porcelain. London: Faber and Faber, 1963. xvi, 35 p. 161 illus. in 100 pls. (4 color). Bibliography, pp. 32-33. Index, pp. 34-35.

Rackham briefly covers the origins and techniques of Italian majolica and its importance as the originator of tin-glazed pottery throughout Europe. Chapters are devoted to its areas of production, development of designs, use of luster, pictorial school, sgraffiato ware, and the later majolica designs.

Roman, Janos. POTTERY OF SAROSPATAK. Budapest: Corvina Press, 1955. 23 p. 80 illus. (part color) in 32 pls. Paper.

This small text surveys the development of the pottery tradition of the town of Sarospatak in northeastern Hungary from the sixteenth to the nineteenth centuries. Photographs of plates, cups, and pitchers illustrate the brightly painted, earthenware designs of this folk tradition.

Ross, Marvin C. RUSSIAN PORCELAINS. Norman: University of Oklahoma Press, 1968. xxviii, 427 p. 86 color pls., 144 halftones, 44 figs. (marks).

Basically a catalog of the porcelains in the collection of the late Marjorie Merriweather Post at Hillwood, Washington, D.C., this text illustrates all of the Russian pieces in that collection except those from the Imperial Porcelain Factory (to be treated in a separate volume). Several pieces from the Rosso collection at Hillwood are also included and examples from some twenty factories, including the Francis Gardner Porcelain Factory, are noted with a brief history and bibliography for each.

*Schmidt, Robert. PORCELAIN AS AN ART AND A MIRROR OF FASHION. Translated by W.A. Thorpe. London: George G. Harrap & Co., 1932. 336 p. 217 illus. including 8 color pls. Bibliography, pp. 311-17. Index, pp. 319-36. (Chamberlin, no. 1759).

The author's stated purpose is to write a brief history of porcelain in its general relation to culture. Treating oriental porcelain only briefly, he focuses on the "golden age" of porcelain, beginning with Augustus the Strong of Saxony and ending with Marie Antoinette.

Solon, M.L. [Louis Marc]. A HISTORY AND DESCRIPTION OF ITALIAN MAJOLICA. London and New York: Cassell & Co., 1907. xvi, 208 p. 73 pls. (24 color). Marks, pp. 195-204. Index, pp. 205-8.

Introducing his subject with an extensive chapter on the nature, development, and manufacture of majolica, Solon devotes individual chapters to each of the centers of majolica manufacture in Italy from the most northern provinces to the Neopolitan states. Each chapter includes a short bibliography and brief chapters at the end of the text are devoted to minor factories and marks.

______. "Old French Porcelain Manufacture in France." Unpublished manuscript, Columbia University, Avery Memorial Library, [190?].

Beginning with a general history of European porcelain manufacture, Solon discusses some nineteen French porcelain centers in individual chapters. No illustrations are included.

Stazzi, Francesco. ITALIAN PORCELAIN. New York: G.P. Putnam's Sons, 1964. 128 p. 135 illus. (42 color).

Intending this volume for the amateur collector and to supplement Landais's FRENCH PORCELAIN with a survey of similar eighteenth-century Italian products, Stazzi includes both rare and fairly average examples from the factories of Florence, Venice, Rome, Naples, and the Piedmont. Although some examples of functional ware are cited, Stazzi's emphasis is on the figures and more decorative pieces.

Van De Put, A[lbert]. THE VALENTIAN STYLES OF HISPANO-MORESQUE POTTERY 1404-1454, ETC. Peninsular Series. New York: Hispanic Society of America, 1938. xi, 99 p. 13 halftones in 7 pls. References, pp. 91-93. Index, pp. 95-99.

Intended as a companion to the APUNTES SOBRE CERAMICA MORISCA of the late G.J. DeOsma, this small volume provides a chronological study of Valentian styles and attempts to identify the various fifteenth-century styles referred to in these apuntes, or documents. Van De Put's text is very dependent upon DeOsma's, which does not appear to have been translated from the Spanish.

Vecht, A. FREDERIK VAN FRYTOM, 1632-1702. Amsterdam: Scheltema & Holkema, 1968. 104 p. 102 illus. (5 color).

This well-illustrated biography of one of the most skilled delft-pottery decorators includes brief chapters on the town of Delft, its pottery and its guild, in its treatment of the life and work of van Frytom. Some ninety-seven pieces of his work are illustrated and documents related to him are included.

Voit, Pal, and Holl, Imre. OLD HUNGARIAN STOVE TILES. Translated by Lili Halapy. Budapest: Corvina Press, 1963. 64 p. 48 pls. (part color). Bibliography, pp. 56-57.

The first part of this small volume discusses the basic development and changing styles of these stove tiles from their beginnings in the fourteenth century to their golden age in the sixteenth century. The second half of the text describes the tile-making workshops of Matthias Corvinus in Buda Castle and of the guilds, mining towns, and border fortresses.

Vydra, Josef, and Kunz, Ludvik. PAINTING ON FOLK CERAMICS. Translated by Roberta Finlayson Samsour. London: Spring Books, n.d. 78 p. 156 pls. (part color). Bibliography, p. 73.

This well-illustrated volume examines the peasant pottery of the Moravian-Slovak people in order to study "the characteristics and thematic repertoire of different potters, their brushwork, the stylistic features of genuine folkart and the models" which served to inspire these native craftsmen rather than to identify the work of outstanding individual craftsmen. Touching upon various themes, the authors devote sections of their text to decoration on majolica, the "Haban" Anabaptists, motifs, Renaissance decoration, brushwork, the styles of the baroque and rococo periods, and the influence of factory styles.

Wallis, Henry. THE ART OF THE PRECURSORS. London: Bernard Quaritch, 1901. 98 p. 94 figs.

Composed principally of illustrations drawn by the author, this text's objective is to illustrate examples of early Italian majolica that had come to light through recent excavations. More contemporary research and the development of superior photographic illustration rather limits this volume's usefulness today.

_____. ITALIAN CERAMIC ART: THE ALBARELLO, A STUDY IN EARLY RENAISSANCE MAJOLICA. London: Bernard Quaritch, 1904. xxix, 117 p. 117 figs.

Accompanying his text with his own illustrations of every known type of albarello, Wallis briefly explains the origin, use, and decoration of these cylindrical, widemouthed, dry storage containers made during the Renaissance and frequently used in _spezierias_ ("pharmacies").

_____. ITALIAN CERAMIC ART: FIGURE DESIGN AND OTHER FORMS OF ORNAMENTATION IN FIFTEENTH CENTURY ITALIAN MAJOLICA. London: Bernard Quaritch, 1905. xxxii, 103 p. 4 color pls., 101 figs.

This last of Wallis's volumes on Italian majolica describes the predominant subjects used by the majolica craftsmen regardless of the object's shape or use, including various religious and allegorical themes, women's portraits, and hunting scenes. The illustrations are from drawings by the author.

_____. ITALIAN CERAMIC ART: THE MAJOLICA PAVEMENT TILES OF THE FIFTEENTH CENTURY. London: Bernard Quaritch, 1902. xxvi, 87 p. 93 figs.

In his introduction Wallis describes the various centers of production and techniques used in the manufacture of these tiles as well as the location of the relatively few surviving examples. The ninety-three tiles illustrated were drawn by the author.

_____. OAK LEAF JARS. London: Bernard Quaritch, 1903. xli, 92 p. 83 figs.

Wallis describes and illustrates a number of fifteenth-century Italian ceramic pieces which can be classified as majolica but which also show a strong Moresco influence. The history of these vases in the nineteenth century and various hypotheses as to their origin are explained in the introduction.

______. THE ORIENTAL INFLUENCE ON THE CERAMIC ART OF THE ITALIAN RENAISSANCE. London: Bernard Quaritch, 1900. xxx, 50 p. 54 figs.

Reasoning that the first potters of the Near East were probably from China and their work retained Far Eastern influences, Wallis draws a number of Near Eastern and Hispano-Moresque ceramic pieces to demonstrate the influence of these styles on the wares of the Italian Renaissance.

Chapter 7
CERAMICS OF THE UNITED STATES, CANADA, AND MEXICO
(American Indian, Canadian, United States, Post-Conquest Mexican)

Where volumes overlap subdivisions, they are generally included under the United States.

AMERICAN INDIAN CERAMICS

Bunzel, Ruth L. THE PUEBLO POTTER: A STUDY IN CREATIVE IMAGINATION IN PRIMITIVE ART. New York: Columbia University Press, 1929. Reprint. New York: Dover Publications, 1972. 134 p. 38 pls. (halftone, fig.). 3 appendixes, pp. 91-128. Bibliography, pp. 129-32. Index, pp. 133-34. Paper.

> Bunzel's text centers around the designs, symbolism, and decorative styles used in Zuni, Acoma, Hopi, and San Ildefonso Pueblo pottery. An extensive selection of designs are illustrated in the appendixes, where the use, name, and meaning of each design motif is briefly described.

Chapman, Kenneth M. THE POTTERY OF THE SAN ILDEFONSO PUEBLO. School of American Research Monograph Series, no. 28. Albuquerque: University of New Mexico Press for the School of American Research, 1970. xvi, 260 p. 174 color pls., 5 figs., map. Bibliography, pp. 256-60.

> A sequel to Chapman's POTTERY OF THE SANTO DOMINGO PUEBLO (below) published in 1936, this large volume includes chapters on the features of San Ildefonso pottery and culinary, black-on-cream, black-on-red, polychrome, and polished wares. The excellent plates illustrate the many decorative motifs used on this pottery and each plate is accompanied by a detailed explanation of the designs. The final chapter, on painted Tewa pottery is by Francis H. Harlow, who completed this volume after Chapman's death in 1968.

______. THE POTTERY OF THE SANTO DOMINGO PUEBLO: A DETAILED STUDY OF ITS DECORATION. Santa Fe, N. Mex.: Laboratory of Anthropology, 1936. 192 p. 34 figs., 79 pls. Bibliography, p. 192.

Although this volume contains some information on the making of Pueblo pottery and its various forms and styles, the major part of its text is devoted to the decoration of this ware and its hundreds of design elements.

_____. PUEBLO INDIAN POTTERY. 2 vols. Nice: C. Szwedzicki, 1936. Each vol.: 16 p., 50 color pls. Portfolio.

These two unusual volumes contain one hundred eleven-by-fourteen-inch color plates of Indian pottery. Volume 1 illustrates the pottery of nine villages in the state of New Mexico and volume 2 deals with Pueblo villages farther to the west.

Curry, Hilda J. NEGATIVE PAINTED POTTERY OF ANGEL MOUNDS SITE AND ITS DISTRIBUTION IN THE NEW WORLD. Indiana University Publications in Anthropology and Linguistics, Memoir no. 5. Baltimore, Md.: Waverly Press, [1950]. 57 p. Illus., maps. Paper.

The forms and styles of negative painted pottery in the southeastern United States (the Tennessee-Cumberland area) are examined and compared with examples from the American Southwest and Central and South America.

Deetz, James. THE DYNAMICS OF STYLISTIC CHANGE IN ARIKARA CERAMICS. Illinois Studies in Anthropology, no. 4. Urbana: University of Illinois Press, 1965. 111 p. Illus., tables. Bibliography, pp. 103-7. Index, pp. 109-11.

This anthropological study seeks to determine the possible relationships between the social structure and ceramics in eighteenth-century Arikara culture. This study used a sample of twenty-five hundred rim shards recovered from the Arikara site in central South Dakota.

Emerson, J. Norman. "Understanding Iroquois Pottery in Ontario: A Rethinking." Rev. ed. Mississauga: Ontario Archaeological Society, 1968. 132 p. Figs., charts, tables. Bibliography, pp. 125-32. Paper. Mimeographed.

Intended as a general introduction for the beginning student of Iroquois ceramics, this guide was originally published in 1956. Pottery shapes and design motifs are clearly treated although the approach is fairly technical and anthropological and suffers from the lack of good illustrations.

*Frank, Larry, and Harlow, Francis H. HISTORIC POTTERY OF THE PUEBLO INDIANS, 1600-1880. Boston: New York Graphic Society, 1974. 160 p. 198 illus. (32 color), map. Glossary, pp. 155-58. Bibliography, pp. 158-60.

Part 1 of this survey identifies the various techniques, forms, and styles of historic Pueblo pottery while part 2 examines examples of different historic types and incorporates a good selection of clear illustrations. Each example illustrated in the second part is accompanied by a detailed caption.

Goddard, Pliny Earle. POTTERY OF THE SOUTHWESTERN INDIANS. 1928. Reprint. New York: American Museum of Natural History, 1971. 30 p. Illus. Bibliography, pp. 12-13. Paper.

This small pamphlet gives a basic introduction to the Indian methods of forming and decorating pottery. Illustrations include photographs of a Zuni potter at work and examples of a variety of forms and decorated ware.

Holmes, William H. "Pottery of the Ancient Pueblos." SMITHSONIAN INSTITUTION, BUREAU OF ETHNOLOGY, ANNUAL REPORT, vol. 4. Washington, D.C.: Smithsonian Institution, 1886. Pp. 257-360. Illus. Index, pp. 523-32.

This is one of four articles published by the Smithsonian on Amerindian pottery for its annual report. The other monographs are: "Ancient Pottery of the Mississippi Valley"; "Form and Ornament in Ceramic Art"; "A Study of Pueblo Pottery as Illustrative of Zuni Culture Growth." The first two of these three are also by Holmes while the third is by Frank H. Cushing.

Kidder, Alfred Vincent. THE POTTERY OF PECOS. Vol. 1: THE DULL-PAINT WARES. Robert S. Peabody Foundation for Archaeology, Papers of the Southwestern Expedition, no. 5. New Haven, Conn.: Yale University Press for Phillips Academy, 1931. 166 p. 102 illus. (1 color). Bibliography, pp. 165-66. Paper.

This general study of the upper Pecos valley in east central New Mexico examines the various styles, forms, and decoration used by the inhabitants of the Forked Lightning ruin and later the Pecos pueblo until its abandonment in 1838. Numerous figures illustrate the various designs used on the pottery and a section on black-on-white wares by Charles Avery Amsden is also included.

Marriott, Alice. MARIA: THE POTTER OF SAN ILDEFONSO.

See: Chapter 8: "Contemporary World Ceramics" (p. 120).

Mera, H[arry]. P[ercival]. STYLE TRENDS OF PUEBLO POTTERY. Memoirs of the Laboratory of Anthropology, vol. 3. Santa Fe, N. Mex.: Laboratory of Anthropology, 1939. 29 p. 67 pls., figs. Bibliography, p. 29. Paper.

Covering wares from the Rio Grande and Little Colorado cultural areas of the sixteenth to nineteenth centuries, this study examines their types and development of forms and decoration. Each plate includes a halftone illustration, a cross section of the piece's contour, a diagrammatic line drawing of its decoration, and a brief description of the form and decoration of the object illustrated.

Morris, Earl H. THE BEGINNINGS OF POTTERY MAKING IN THE SAN JUAN AREA: UNFIRED PROTOTYPES AND THE WARES OF THE EARLIEST CERAMIC PERIOD. Anthropological Papers, vol. 28, pt. 2. New York: American Museum of Natural History, 1927. 73 p. Illus., map. Paper.

Morris attempts to trace the origins and development of the earliest ceramic types in the San Juan drainage area east of the Chin Lee valley in Arizona. In addition to describing periods and vessel types, he explains several traditional pottery designs.

Morss, Noel. CLAY FIGURINES OF THE AMERICAN SOUTHWEST. Papers of the Peabody Museum of Archaeology and Ethnology, Harvard University, vol. 49, no. 1. Cambridge, Mass.: Peabody Museum, 1954. x, 75 p. 31 halftones, figs., 2 maps. References, pp. 67-73. Paper.

Primarily intending to illustrate and describe a series of eleven important clay figurines discovered in Range Creek, Utah, in 1950, Morss also reviews the cultural and chronological position of figurines from both the northern and southern Indian cultural traditions. He concludes with a discussion of the purpose of the figures and their significance to their makers.

New York. Museum of the American Indian. NAKED CLAY: 3000 YEARS OF UNADORNED POTTERY OF THE AMERICAN INDIAN. New York: Heye Foundation, 1972. 76 p. 107 halftones. Bibliography, pp. 73-76. Paper.

This exhibition catalog includes an essay by Lewis Krevolin on the technology of Amerindian pottery and ninety examples of pottery from North and South America. These examples are annotated and display an extensive range of creative forms.

Putnam, F[rederic].W[ard]. "Pueblo Pottery." AMERICAN ART REVIEW, February 1881, pp. 150-54. 1 pl., 7 figs.

Putnam's article describes several pieces of "modern" Pueblo pottery and compares their decoration with much older examples of Pueblo wares. Some information on the materials and technical skills used in making these wares is also mentioned.

Stiles, Helen E. POTTERY OF THE AMERICAN INDIANS. New York: E.P. Dutton & Co., 1939. 169 p. Halftones. Bibliography, pp. 161-65. Index, pp. 166-69.

Informally written as a general introduction to the subject, Stiles's text discusses Indian pottery in northern America, the Southwest, prehistoric Peru, the Yucatan peninsula, and past and present

Mexico. Illustrations survey several examples from each of these areas.

Tschopik, Harry, Jr. NAVAHO POTTERY MAKING. Papers of the Peabody Museum of Archaeology and Ethnology, Harvard University, vol. 17, no. 1. Cambridge, Mass.: Peabody Museum, 1941. viii, 85 p. 16 pls. Bibliography, pp. 77-79. Paper.

Desiring to make a detailed study of observed field material before it disappears, Tschopik describes the technological details of pottery manufacture and supplements them with interviews. He also attempts to relate the making of Navaho pottery, particularly the manufacture of ritual pipes, to its cultural and social context and to compare it with the pottery making of other southwestern groups.

Wormington, H[annah].M[arie]., and Neal, Arminta. THE STORY OF PUEBLO POTTERY. Denver Museum of Natural History Pictorial, no. 2. Denver: Denver Museum of Natural History, 1951. 61 p. Halftones, maps. Bibliography, p. 60. Paper.

This museum pamphlet briefly outlines the history of Pueblo pottery from its beginnings with the Basket Makers through its decline. A good number of chronologically arranged halftones of both ancient and contemporary pieces supplement the text.

CANADIAN CERAMICS

Collard, Elizabeth. NINETEENTH-CENTURY POTTERY AND PORCELAIN IN CANADA. Montreal: McGill University Press, 1967. xx, 441 p. 56 pls. 2 appendixes, pp. 323-50. Index, pp. 431-41.

Collard examines imported porcelain, earthenware, and printed earthenware with a Canadian view because imported wares constituted such an extensive part of Canadian pottery in the nineteenth century. About one third of this history is devoted to native potters and their work. The two appendixes contain marks and a checklist of nineteenth-century Canadian potters.

Webster, Donald Blake. THE BRANTFORD POTTERY. Art and Archaeology, Occasional Papers, no. 13. Toronto: Royal Ontario Museum, 1968. 80 p. 78 illus., maps. Bibliography, p. 72. Paper.

From historical records, excavations, and recoveries, Webster describes the history of this pottery (1849-1907) and assesses its products, particularly the range of middle-period (1873-83) yellow and Rockingham-glazed wares, which had not previously been attributed because they lacked markings.

_____. EARLY CANADIAN POTTERY. Greenwich, Conn.: New York Graphic Society, 1971. 256 p. Illus. (8 color). Glossary, pp. 247-48. Bibliography, pp. 249-50. Index, pp. 251-53.

Written by the curator of the Canadiana department of the Royal Ontario Museum, this profusely illustrated volume covers both the coarse earthenwares and the mechanically formed industrial pottery of Canada. The book outlines the range of early Canadian pottery rather than providing a history of all its potters and ends with a chapter on the archaeology of the potteries.

_____. EARLY SLIP-DECORATED POTTERY OF CANADA. Toronto: Charles J. Musson, 1969. 75 p. 55 halftones. Bibliography. Index. Paper.

Intended as an informal study for the use of collectors, this text surveys the types, origins, and periods of the slipwares found in Canada. Seventeenth-, eighteenth-, and nineteenth-century examples of native and imported wares are included.

U.S. CERAMICS

Altman, Violet, and Altman, Seymour. THE BOOK OF BUFFALO POTTERY. New York: Crown Publishers, Bonanza Books, 1969. 192 p. 378 illus. (8 color). Glossary, pp. 184-86. Bibliography, p. 187. Index, pp. 188-92.

This well-illustrated history of the Buffalo Pottery factory explains its connection with the Larkin Company of Buffalo and the various styles of pottery which the former produced, including the famous Deldare ware.

American Crafts Council. FORMS FROM THE EARTH: 1,000 YEARS OF POTTERY IN AMERICA. New York: 1962. 21 p. 25 loose pls. Paper.

This catalog of American ceramic history is broken down into four parts--the American Indian, the colonial period to the Civil War, the Civil War to the turn of the century, and the twentieth century. Each part begins with a brief essay by a different author and the catalog section identifies each of the 300 items exhibited. Each of the loose plates is a 7 1/2-by-9-inch halftone.

Barber, Edwin Atlee. ANGLO-AMERICAN POTTERY: OLD ENGLISH CHINA WITH AMERICAN VIEWS. 2d ed. Philadelphia: Patterson & White Co., 1901. 220 p. Illus. 2 indexes, pp. 205-20.

Containing information on designs created by Liverpool, Staffordshire, and unknown potters, this source illustrates a large selection of English china made for the American market. Also included is a chapter on ceramic printing.

_____. THE POTTERY AND PORCELAIN OF THE UNITED STATES: AN HISTORICAL REVIEW OF AMERICAN CERAMIC ART FROM THE EARLIEST TIMES TO THE PRESENT DAY. 3d ed. rev. New York: G.P. Putnam's Sons, 1909. xxviii, 621 p. 335 illus. Index, pp. 601-21. (Chamberlin, no. 1780).

This old volume touches on a diverse range of material in the history of American ceramics, including American china manufacturing, tobacco pipes, ornamental tiles, and the potteries of Mexico.

_____. SALT GLAZED STONEWARE: GERMANY, FLANDERS, ENGLAND AND THE UNITED STATES.

See: Chapter 6: "Western Ceramics: The Middle Ages to the Twentieth Century" (p. 88).

_____. TULIP WARE OF THE PENNSYLVANIA-GERMAN POTTERS: AN HISTORICAL SKETCH OF THE ART OF SLIP-DECORATION IN THE UNITED STATES. Philadelphia: Pennsylvania Museum, 1903. 233 p. 94 illus. Index, pp. 225-33. Paper.

This volume contains a variety of information on the history and techniques of American slip-decoration, including chapters on local dialect, inscriptions, and decorative subjects.

Barret, Richard Carter. BENNINGTON POTTERY & PORCELAIN: A GUIDE TO IDENTIFICATION. New York: Bonanza Books, 1958. 348 p. 457 illus. (7 color). List of illustrations, pp. 343-45. Index of illustrations, pp. 346-48.

This volume includes a brief history of and illustrates the various types of items and wares produced at Bennington during the middle of the nineteenth century. Divided into functional items and fancy articles, every known type of ware and all of the most significant production items are illustrated.

Benjamin, Marcus. AMERICAN ART POTTERY. Reprint. Washington, D.C.: Reprinted from THE GLASS & POTTERY WORLD, 1907. 57 p. Illus.

This small volume is a report on the manufacture of American art pottery based on Benjamin's visits to several U.S. centers, including New Jersey, Ohio, West Virginia, and Boston. He gives a brief history of the developing industry beginning in the 1870s, describing manufacturing techniques and the types of wares produced.

Beurdeley, Michel. CHINESE TRADE PORCELAIN.

See: Chapter 4: "Eastern Ceramics" (p. 27).

Bivins, John, Jr. THE MORAVIAN POTTERS IN NORTH CAROLINA. Chapel Hill: University of North Carolina Press, 1972. 301 p. 278 illus. (2 color). Glossary, pp. 279-83. Glossary of Moravian terms, pp. 285-86. Bibliography, pp. 289-90. Index, pp. 291-97.

This history examines the earthenware pottery of the Moravian religious sect in the area of Salem, North Carolina, giving particular attention to the eighteenth-century craftsmen Gottfried Aust, Rudolf Christ, and Gottlob Krause.

Christensen, Erwin O. EARLY AMERICAN DESIGNS: CERAMICS. New York: Pitman Publishing Corp., 1952. 49 p. Figs. Index, p. 49. Paper.

Its only text a two-page foreword by Christensen, this volume is composed of a selection of designs reproduced from the INDEX OF AMERICAN DESIGN at the National Gallery of Art, Washington, D.C. All of the design examples are from pieces produced in Ohio and the American Northeast.

Clarke, John M. THE SWISS INFLUENCE ON THE EARLY PENNSYLVANIA SLIP DECORATED MAJOLICA. Albany, N.Y.: J.B. Lyon Co., 1908. 18 p. 8 halftones.

Clarke, director of the New York State Museum, simply distinguishes a group of slip-decorated pieces from similar, German-inspired work, comparing their motifs to the Swiss originals.

Clement, Arthur W. OUR PIONEER POTTERS. York, Pa.: Maple Press, 1947. ix, 94 p. 24 halftones in 15 pls. Bibliography, pp. 89-91. Index, pp. 93-94.

Not intending to provide a complete list of American potters and potteries of the past 300 years, Clement discusses in some detail the work of a relatively small number of exceptionally interesting potteries. Generally dividing the latter by types of wares--red, stone, molded, utility, and decorative wares and porcelains--Clement describes the various artists and firms which he has selected, giving his text authority by frequent documentation of his sources.

Crawford, Jean. JUGTOWN POTTERY: HISTORY AND DESIGN. Winston-Salem, N.C.: John F. Blair, Publisher, 1964. 127 p. 25 illus. (12 color). Bibliography, pp. 120-27.

This volume is divided into three parts: a history of Jugtown pottery, its composition and design, and a look at the future of this pottery tradition; the first two parts make up the bulk of the book. The whole text is a tribute to the efforts of Jacques and Juliana Busbee for their years of effort at revitalizing the pottery industry at Jugtown.

Earle, Alice Morse. CHINA COLLECTING IN AMERICA. New York: Charles Scribner's Sons, 1892. 429 p. 66 figs. Index, pp. 425-49.

Intended as a collector's guidebook, Earle's text devotes individual chapters to china made in America or for the American market.

Several chapters describe Washington and presidential china and designs relating to Franklin, Lafayette, and other patriotic subjects. The illustrations are small and not very clear.

Evans, Paul. ART POTTERY OF THE UNITED STATES: AN ENCYCLOPEDIA OF PRODUCERS AND THEIR MARKS. New York: Charles Scribner's Sons, 1974. 353 p. Illus. (8 color), marks. 4 appendixes, pp. 343-45. Index, pp. 346-52.

Evans states three clear purposes for his encyclopedia: to establish an accurate history of the various art pottery operations in the United States emphasizing the individuals, not the plants involved, nor the artistic influences on these individual works; to provide a comprehensive listing of marks; and to indicate where examples by a particular potter, artist, or factory can be studied. A number of the illustrations here are vintage photographs of artists, potteries, exhibitions, and studio interiors.

Greer, Georgeanna H., and Black, Harding. THE MEYER FAMILY: MASTER POTTERS OF TEXAS. San Antonio, Tex.: San Antonio Museum Association, 1971. 97 p. Illus. (4 color). Paper.

Published in conjunction with an exhibition of Meyer pottery held at the Witte Memorial Museum, San Antonio, Texas, this study recounts the history of the Meyer pottery from 1887 to 1964. A good selection of pieces produced during the different years of the pottery's existence is illustrated and described.

Guilland, Harold F. EARLY AMERICAN FOLK POTTERY. Philadelphia: Chilton Book Co., 1971. 322 p. Illus. (16 color), map. Bibliography, pp. 293-309. Index, pp. 311-22.

Beginning with a history of the folk pottery tradition as a whole in the United States, this excellent volume is for the most part, composed of a large number of unique illustrations. These illustrations are renderings, not photographs, of folk pottery selected from the INDEX OF AMERICAN DESIGN begun in the 1930s.

*Henzke, Lucile. AMERICAN ART POTTERY. Camden, N.J.: Thomas Nelson, 1970. 336 p. Illus. (16 color), marks. Glossary, pp. 323-25. 2 indexes, pp. 326-36.

Illustrated with hundreds of pottery examples, marks, and signatures, this volume describes the products of thirty or more nineteenth- and twentieth-century pottery companies in the United States. Besides the more well-known art potteries, Henzke also includes several tile companies, the Jugtown pottery, and the work of Maria Martinez.

Hillier, Bevis. POTTERY AND PORCELAIN 1700-1914: ENGLAND, EUROPE AND NORTH AMERICA.

See: Chapter 6: "Western Ceramics: The Middle Ages to the Twentieth Century" (p. 90).

Hood, Graham. BONNIN & MORRIS OF PHILADELPHIA. Institute of Early American History and Culture at Williamsburg, Va. Chapel Hill, N.C.: University of North Carolina Press, 1972. xiii, 78 p. 56 halftones in 32 pls. 15 appendixes, pp. 47-68. Index, pp. 77-78.

This history of the first American porcelain factory (1770-72), in Philadelphia, includes documentary information from records, analyses of the nature of the wares themselves, and evidence from site excavations at the factory.

Hyde, J.A. Lloyd. ORIENTAL LOWESTOFT. New York: Charles Scribner's Sons, 1936. 161 p. 31 pls. (1 color). Index, pp. 158-61.

Placing special emphasis on the China trade and porcelain decorated specifically for the American market, the author defines this ware (oriental hard-paste porcelain ordered by Europeans and Americans through agents in Canton), explains the origins of the trade, and the role of the East Indies companies. Additional chapters are devoted to types of decoration, armorial pieces, and pieces made specifically for the American market.

James, Arthur E. THE POTTERS AND POTTERIES OF CHESTER COUNTY, PENNSYLVANIA. [West Chester, Pa.]: Chester County Historical Society, 1945. 116 p. 20 illus., map. List of potters, pp. 111-14. Bibliography, pp. 115-16.

This brief history gives information on the materials and tools used, the types of wares produced, and the families involved in pottery production in Chester County.

Ketchum, William C., Jr. EARLY POTTERS AND POTTERIES OF NEW YORK STATE. New York: Funk & Wagnalls, 1970. 278 p. 18 halftones, map. List of New York potters and their marks, pp. 203-51. Bibliography, pp. 254-69. Index, pp. 270-78.

Although it has very few illustrations, this interesting volume does contain a great deal of historical information on literally hundreds of potteries in New York State from the seventeenth to the nineteenth centuries.

_____. THE POTTERY AND PORCELAIN COLLECTOR'S HANDBOOK: A GUIDE TO EARLY AMERICAN CERAMICS FROM MAINE TO CALIFORNIA. New York: Funk & Wagnalls, 1971. xx, 204 p. 36 halftones in 16 pls. Appendix, pp. 159-93. Index, pp. 195-204.

Dividing his text into five general sections--redware, stoneware, yellowware, whiteware, and porcelain--Ketchum begins each section with a description of relevant manufacturing techniques and identifies numerous factories, artists, studios, and types of wares produced in the different geographic areas and local centers across the country. The illustrations are limited, but hundreds of names are identified and an extensive list of early potteries is included in the appendix.

*Kovel, Ralph, and Kovel, Terry. THE KOVELS' COLLECTOR'S GUIDE TO AMERICAN ART POTTERY. New York: Crown Publishers, 1974. 378 p. 16 color pls. Index, pp. 369-78.

Arranged alphabetically by factory, this guide covers the production of art pottery in the United States from the 1870s to the 1920s and presents brief sections on pottery tiles and miscellaneous factories at the end. Each factory entry contains information on makers, artists, dates, marks, lines of ware, a general history of the pottery, and a brief bibliography.

Laidacker, Sam, ed. THE STANDARD CATALOGUE OF ANGLO-AMERICAN CHINA: FROM 1810 TO 1850. Scranton, Pa.: Sam Laidacker, 1938. 91 p. 57 illus. (Chamberlin, no. 1781).

This catalog identifies a variety of printed and decorated ware made in England by some dozen Staffordshire companies for the American trade. Including information on the makers, border patterns, views, colors, and marks, the catalog also estimates the market price of the particular pieces in "proof" condition at the time of the catalog's publication.

Low, J[ohn]., and Low, J.C. ILLUSTRATED CATALOGUE OF ART TITLES. Boston: C.A. Wellington & Co., 1881. 5 p. 26 pls.

This catalog advertises and illustrates an extensive number of art tile designs by John Low for fireplaces and various architectural features. The designs of individual tiles and complete sets are illustrated.

McCauley, Robert H. LIVERPOOL TRANSFER DESIGNS ON ANGLO-AMERICAN POTTERY. Portland, Me.: Southworth-Anthoensen Press, 1942. xxi, 150 p. 32 pls. 113 halftones. Bibliography, p. 141. List of Liverpool potters, p. 142. Index, pp. 143-50.

McCauley lists some 265 transfer designs used on Liverpool pottery destined for the American trade. The designs fall into four groups: portraits of prominent Americans; depictions of events in American military and naval history; scenes of American patriotism, trade, and industry; and representations of particular localities. Each design is briefly described, but not every design is illustrated.

Mudge, Jean McClure. CHINESE EXPORT PORCELAIN FOR THE AMERICAN TRADE, 1785-1835. A Winterthur Series Book. [Newark]: University of

Delaware Press, 1962. 284 p. 138 halftones. Notes, pp. 197-224. 2 appendixes, pp. 227-32. Bibliography, pp. 235-49. Index, pp. 253-84.

Part 1 of this extensively documented study briefly explains China's relationship with Europe and America and includes a description of the trade at Canton. The larger part of this work examines the porcelain, its manufacture, sale, and shipment, as well as the decoration and chronology of the wares exported to America.

Munson-Williams-Proctor Institute. WHITE'S UTICA POTTERY. Utica, N.Y.: 1969. Unpaged. 10 illus. Paper.

This catalog of a historical exhibition of pottery manufactured in Utica, New York includes a brief introduction to the White Pottery by Barbara Franco, curator of decorative arts at the Munson-Williams-Proctor Institute.

New Jersey State Museum. NEW JERSEY POTTERY TO 1840. Trenton: 1972. Unpaged. Illus. Bibliography. Paper.

This exhibition catalog illustrates a number of examples of lead-glazed redware, salt-glazed stoneware, and yellow-glazed stoneware and provides a brief textual introduction to these wares and to the several local pottery centers where they were produced.

New York. State University of, at New Paltz. College Art Gallery. AN EXHIBITION OF 18TH AND 19TH CENTURY AMERICAN FOLK ART. New Paltz: 1969. 35 p. 46 halftones. Paper.

Beginning with a two-page introduction to American folk pottery by Richard I. Barons, this exhibition catalog describes each of the sixty-three pieces in the exhibition. The objects come from New England, New York, Pennsylvania, Ohio, and the South. About two-thirds of the pieces are illustrated.

New York City. Metropolitan Museum of Art. A MEMORIAL EXHIBITION OF PORCELAIN AND STONEWARE BY ADELAIDE ALSOP ROBINEAU, 1865-1929. New York: 1929. 24 p. 2 halftones. Paper.

This small catalog contains Joseph Breck's brief introduction to Robineau's life and work and lists the seventy-one pieces assembled for this memorial exhibition of her porcelain and stoneware.

New York City. Museum of American Folk Art. THE POTTERY OF THE STATE. New York: 1974. 14 p. Illus. Paper.

Pertaining to stonewares and redwares made at pottery centers throughout New York State, this exhibition catalog documents the third of five exhibitions slated by the museum in honor of the bicentennial celebration. William C. Ketchum, Jr.'s text briefly outlines the major developments in New York's pottery industry from the seventeenth century until its virtual cessation at the end of the nineteenth century.

Noel-Hume, Ivor. POTTERY AND PORCELAIN IN COLONIAL WILLIAMSBURG'S ARCHAEOLOGICAL COLLECTIONS. Colonial Williamsburg Archaeological Series, no. 2. Williamsburg, Va.: Colonial Williamsburg, 1969. 46 p. 43 halftones. Paper.

This pamphlet illustrates and identifies a variety of domestic and imported ceramic wares excavated at Williamsburg. Examples of stoneware, earthenware, English delft, salt-glazed ware, creamware and exported oriental porcelain are described.

R.W. Norton Art Gallery. THE AMERICAN PORCELAIN TRADITION: 18TH, 19TH AND 20TH CENTURIES. Shreveport, La.: 1972. 32 p. 19 illus. (5 color). Bibliography, pp. 29-30. Index, pp. 31-32. Paper.

This catalog of a loan exhibition from the New Jersey State Museum includes a brief history of porcelain, with special emphasis on its production in America, and identifies each of the seventy post-1830 objects in the exhibition, illustrating several of them.

Osgood, Cornelius. THE JUG, AND RELATED STONEWARE OF BENNINGTON. Rutland, Vt.: Charles E. Tuttle Co., 1971. 222 p. 31 pls. (8 color), 38 figs. Glossary, pp. 185-88. Bibliography, pp. 207-12. Index, pp. 217-22.

Written by a professor of anthropology at Yale University, this interesting volume covers the one-hundred-year history of salt-glaze pottery at the Norton potteries in Bennington, Vermont, and includes brief chapters on the history of pottery and the ceramic process.

Pitkin, Albert Hastings. EARLY AMERICAN FOLK POTTERY: INCLUDING THE HISTORY OF THE BENNINGTON POTTERY. Hartford, Conn.: Mrs. Albert Hastings Pitkin, 1918. 152 p. 26 halftones. Index, pp. 146-52.

Published by the author's widow, this volume is divided into two parts. Part 1 gives a history of the Bennington Pottery, a list of its marks, and a catalog of Pitkin's collection of its wares in the Morgan Memorial, Hartford; part 2 provides a more general history of early American folk pottery, from the area between Maine and Pennsylvania, and also includes a catalog of Pitkin's collection in the Morgan Memorial.

Purviance, Louise; Purviance Evan; and Schneider, Norris. ZANESVILLE ART POTTERY IN COLOR. Leon, Iowa: Mid-America Book Co., 1968. 50 p. 23 color pls. Paper.

This spiral-bound collector's manual illustrates the art wares produced by several pottery companies in the Zanesville, Ohio, area and gives histories and product descriptions of the Weller, Owens, Roseville, Mosaic Tile, and American Encaustic companies. Each color plate illustrates nine to twelve ceramic pieces and there is some information on marks.

Quimby, Ian M.G., ed. CERAMICS IN AMERICA. Winterthur Conference Report, 1972. Charlottesville: University Press of Virginia for the Winterthur Museum, 1973. Illus. Paper.

This collection of papers delivered at the eighteenth Winterthur Conference, on the relationship of ceramic objects to social history, includes studies of remains found at Plymouth, Massachusetts; St. Mary's City, Maryland; Fortress Louisbourg; and Spanish colonial sites. English and American ceramic manufacturing are also discussed in studies of English creamware, pearlware, and salt-glazed stoneware; and American earthenware, stoneware, and porcelain.

Ramsay, John. AMERICAN POTTERS AND POTTERY. [Boston]: Hale, Cushman & Flint, 1939. 304 p. 91 halftones, drwgs. Bibliography, pp. 244-51. Potter's marks, pp. 252-85. Indexes, pp. 287-304.

Written from the standpoint of the collector and with an emphasis on "collectable" rather than on early or recent pottery, Ramsay's book deals with the pottery of the East Coast, the South, and the Midwest. Also included is an eighty-page checklist of American potters.

Rice, A[lvin].H., and Stoudt, John Baer. THE SHENANDOAH POTTERY. Strasburg, Va.: Shenandoah Publishing House, 1929. 277 p. 45 pls. (7 color).

The first part of this volume is a short history of several nineteenth-century potteries in the Shenandoah Valley; the second part a catalog of over 2,000 pieces of Shenandoah pottery in the Rice collection. Some one hundred pieces in the collection are illustrated here.

Rochester Museum and Science Center. CLAY IN THE HANDS OF THE POTTER. Rochester, N.Y.: 1974. 56 p. 57 illus. Bibliography, p. 56. Paper.

This exhibition catalog of pottery manufactured in the Rochester and Genesee valley region during 1793-1900 includes introductory essays on the region's earthenwares and stonewares and brief histories of some eleven local potteries. Forty-eight examples are identified and illustrated.

*Schwartz, Marvin D., and Wolfe, Richard. A HISTORY OF AMERICAN ART PORCELAIN. New York: Renaissance Editions, 1967. 93 p. 75 pls. (7 color). Bibliography, p. 93.

Combining a great many illustrations with a very concise text, this popular history briefly surveys the development of American manufactured porcelain from the middle of the eighteenth century to the present.

Smith, Elmer L. POTTERY: A UTILITARIAN FOLK CRAFT. Lebanon, Pa.: Applied Arts Publishers, 1972. 32 p. Halftones. Paper.

This booklet deals with the products of the later stages--1830 to the turn of the century--of American pottery making and briefly describes the common wares produced by some dozen inland frontier potteries. A good number of stoneware and earthenware pieces are illustrated.

Snively, R[obley].D[unglison]., and Snively, M[ary].E[lizabeth]. POTTERY. Old Deerfield Series of Handicraft Manuals. Brattlesboro, Vt.: Stephen Daye Press, 1940. 87 p. Illus.

This book briefly and informally portrays the potter's art through photographs of Edward d'Ambigny Norman working at the potter's wheel.

Spargo, John. EARLY AMERICAN POTTERY AND CHINA. Century Co., 1926. Reprint. Garden City, N.Y.: Garden City Publishing Co., 1948. xvii, 393 p. 64 pls. Appendix of marks, pp. 357-72. Bibliography, pp. 373-76. Index, pp. 377-93.

Although Spargo modestly states that this volume is "by a hobbyist for his fellow hobbyists," his work contains more than a small share of very valuable information. Chapters are devoted to classification, types of ware, potters before and after the American revolution, and early and late nineteenth-century potteries. Various lists of potters, potteries, types of ware, and marks are also included, as well as a final chapter entitled "Grotesqueries, Satires & Jests," which illustrates and describes a number of the more unique examples of the potter's art.

______. THE POTTERS AND POTTERIES OF BENNINGTON. Boston: Houghton Mifflin Co., 1926. Reprint. New York: Dover Publications, 1972. xv, 270 p. 44 pls. (8 color). 5 appendixes, pp. 247-55. Index, pp. 257-65.

Illustrated with pieces from his own collection, Spargo's study is divided into two parts, one covering the Norton potteries (1793-1894), the second covering the Fenton. The histories of both potteries are treated in some detail and the identification of marks is included. The appendixes contain information on the formulas of Bennington clay bodies and an extract on the geology of Vermont.

Stiles, Helen E. POTTERY IN THE UNITED STATES. New York: E.P. Dutton & Co., 1941. 329 p. Illus. Bibliography, pp. 317-24. Index, pp. 325-29.

Parts of this wide-ranging text are devoted to pottery made in various areas of the United States, antique pottery, art wares, studio potters, tile, and architectural terra-cotta. Although much of this then contemporary work is now dated, the large selection of illustrations and the varying quality of the work illustrated make interesting viewing some twenty-five years after publication.

Watkins, C. Malcolm. NORTH DEVON POTTERY AND ITS EXPORT TO AMERICA IN THE 17TH CENTURY.

See: Chapter 6: "Western Ceramics: The Middle Ages to the Twentieth Century" (p. 85).

Watkins, Lura Woodside. EARLY NEW ENGLAND POTTERS AND THEIR WARES. Hamden, Conn.: Archon Books, 1950. 291 p. 136 illus. Check list of New England potters, pp. 249-70. Bibliography, pp. 271-76. Index, pp. 283-91.

Although material on nineteenth-century art pottery and industrially produced wares is included, this history emphasizes the common red earthenware and later stoneware produced by the traditional potters in New England from the seventeenth century onward. Many pottery locales throughout New England are covered and a checklist of several hundred New England potters is included.

Webster, Donald Blake. DECORATED STONEWARE POTTERY OF NORTH AMERICA. Rutland, Vt.: Charles E. Tuttle Co., 1971. 232 p. 300 illus. Checklist of potteries, pp. 215-26. Glossary, pp. 227-28. Bibliography, pp. 229-30. Index, pp. 231-32.

Essentially dealing with North American salt-glazed stonewares, Webster's text examines their types, forms, areas of origin, decoration, and European antecedents. Basically, however, the text is organized according to decorative subject matter.

Wiltshire, William E. III. FOLK POTTERY OF THE SHENANDOAH VALLEY. New York: E.P. Dutton Co., 1975. 127 p. 60 color pls., 2 maps. Appendix, pp. 124-27. Bibliography, p. 127.

Although it does include a brief history of the potters and types of wares produced in the Shenandoah valley, this book consists almost entirely of a series of some sixty large color illustrations of these wares, each with a brief, identifying caption.

POST-CONQUEST MEXICAN CERAMICS

Barber, Edwin Atlee. THE MAJOLICA OF MEXICO. Philadelphia: Pennsylvania Museum and School of Industrial Art, 1908. 115 p. 61 illus. Index, pp. 109-115.

Barber examines the history of majolica production in Mexico, its origins in Spain, and its relationship to tin-enameled pottery. Brief chapters on pottery marks and forgeries are included.

Chapter 8
CONTEMPORARY WORLD CERAMICS
(U.S. Ceramics, International Ceramics)

The majority of these works are exhibition catalogs of one-artist, student, and invitational shows. If an exhibition deals exclusively with one artist, I have entered its catalog alphabetically under the artist's surname, whether or not that name appears in the actual title. As I have done throughout this bibliography, I have cataloged all other exhibitions alphabetically by the name of the author of the catalog text or the name of the institution that organized the exhibition.

Additional information on contemporary ceramic catalogs can be found in THE WORLDWIDE ART CATALOGUE BULLETIN (see chapter 1, p. 5).

U.S. CERAMICS

Andreson, Laura. CERAMICS--FORM AND TECHNIQUE. Los Angeles: University of California, Art Galleries, 1970. 24 p. 31 halftones. Paper.

This catalog of a retrospective exhibition of 133 pieces of Andreson's work includes a brief essay by J. Bernard Kester.

Bohrod, Aaron. A POTTERY SKETCHBOOK. Madison: University of Wisconsin Press, 1959. 196 p. Illus.

Except for a brief preface, this volume is composed of a selection of Bohrod's drawings from four of his pottery sketchbooks and of a few photographs of the decorated surfaces of several of his pieces.

Boston. Society of Arts and Crafts. CERAMICS BY SIX. Boston: 1966. 6 p. 6 halftones. Paper.

An introductory essay by Ben Watkins describes the work of these "new forms" artists, Stephen Kaltenbach, Michael Frimkess, Joseph Pugliese, Richard Shaw, David MacKenzie, and James Melchert. Each artist is represented by one illustration.

Brand Library Art Center. BRAND IV CERAMIC CONJUNCTION. Glendale, Calif.: 1974. 8 p. 13 loose pls. Paper.

This exhibition catalog includes a complete list of the participating artists, loose plates of thirteen objects in the exhibition, and an introductory essay entitled "Ceramics as Mainstream Art" by Clark Polak. All of the pieces illustrated are sculptural rather than functional.

_____. BRAND V CERAMIC CONJUNCTION. Glendale, Calif.: 1975. 24 p. 16 halftones. Paper.

This catalog contains a brief introduction to "high art" ceramics, a list of invited and juried entries, and illustrations of sixteen of the objects entered.

Brown, Charles M. RETROSPECTIVE 1951-1969. [Jacksonville, Fla.]: Jacksonville Art Museum, 1969. 12 p. 12 halftones. Paper.

This small catalog lists the ninety-three pieces in the exhibition, illustrates twelve of them, and provides a resume of Brown's work.

California. University of, at Irvine. Art Gallery. ABSTRACT EXPRESSIONIST CERAMICS. Irvine, Calif.: 1966. 54 p. Illus. (8 color). Paper.

This catalog documents an exhibition of the work of ten American ceramic artists, including Billy Al Bengston, John Mason, Kenneth Price, Henry Takemoto, and Peter Voulkos.

California. State University of, at San Francisco. Hayward Art Gallery. NUT ART. San Francisco: 1972. 35 p. 24 illus. Bibliography. Paper.

This exhibition catalog of "new form" ceramic art includes statements by the eighteen participating artists, photographs of them, and illustrations of their work. Some of the works illustrated are ceramic pieces; others are sketches and drawings.

Collier, William Lewis. "Representative Potters of Northern California: Their Work and Philosophy." Master's thesis, University of California, 1956. 92 p. Bibliography, pp. 32-34. Photocopy.

Collier interviews a number of potters and examines their self-set standards for their work and the various traditions upon which they draw in forming that work.

A DECADE OF CERAMIC ART 1962-1972: FROM THE COLLECTION OF PROFESSOR AND MRS. R. JOSEPH MONSEN.

See: Chapter 9: "Ceramic Collections" (p. 141).

DeStaebler, Steven. STEVEN DESTAEBLER SCULPTURE. [Oakland, Calif.]: Oakland Museum Art Gallery, 1974. 32 p. 41 illus. (part color). Paper.

Harvey L. Jones, curator of this exhibition, introduces the catalog, which lists the thirty-nine pieces exhibited and illustrates them as works in progress and as completed sculptures.

Everson Museum of Art of Syracuse and Onandaga County. 25TH CERAMIC NATIONAL EXHIBITION, (1968-1970). Syracuse, N.Y.: 1968. 48 p. Halftones. Paper.

The catalog lists each of the 288 pieces in the exhibition and includes comments by several participating artists on the role of the Ceramic National in the field of the ceramic arts. Started in 1930 as the Adelaide Robineau Memorial Exhibition, this exhibition gradually developed into a national ceramic competition. Held biennially since 1945, the Everson Museum discontinued its biennial show as a national competition after 1968, although it still continued to sponsor a biennial ceramic exhibition. Its 1976 ceramic exhibition was simply entitled CLAY. Catalogs were issued for each of the Everson Museum exhibitions and a second catalog was issued for several of the ceramic national traveling exhibitions.

Fort Wayne Museum of Art. FORT WAYNE INVITATIONAL CERAMICS EXHIBITION. Fort Wayne, Ind.: 1968. 19 loose p. 17 illus. Paper.

Each plate in this catalog is devoted to one of the seventeen artists invited to the exhibition and includes a brief biography and an illustration of the individual's work. Unfortunately, the catalog is printed on a dark tan paper that ruins much of the clarity of the illustrations.

Grotell, Maija. MAIJA GROTELL. New York: Museum of Contemporary Crafts, 1968. 14 p. 11 halftones. Paper.

This catalog lists the sixty-five pieces assembled for this retrospective and compiles biographical information on the artist as well as quotations by her on her work.

Harlow, Francis H., and Young, John V. CONTEMPORARY PUEBLO INDIAN POTTERY. Santa Fe: Museum of New Mexico, 1965. 24 p. 13 illus. (5 color). Paper.

While neither a history of Pueblo pottery nor a technical manual, this pamphlet briefly examines the types of Pueblo pottery most commonly offered for sale, their sources, and methods of production.

Iowa State University. The New Gallery. CLAY TODAY. N.p.: 1962. 20 p. 25 loose pls. Paper.

This catalog begins with brief introductions by James McKinnell and Abner Jonas and includes the professional backgrounds of the fifty-odd artists represented in the exhibit. Two or three objects are illustrated on each of the loose plates.

Kentucky. University of. "A WILD MOOSE CHASE." Lexington: 1973. 16 p. Halftones. Paper.

This catalog of a student-faculty exhibition of ceramics at the University of Kentucky includes a brief biography, portrait photograph, statement, and reproduction of a work of each of the sixteen participants in the exhibition.

Kudlacek, John. SABBATICAL EXHIBITIONS: "A CERAMIC GARDEN." Emporia: Kansas State Teachers College Art Gallery, 1971. 11 halftones. Paper.

This catalog consists solely of eleven loose halftone plates of environmental ceramic pieces.

Lewis, Lucy M. A TRIBUTE TO LUCY M. LEWIS: ACOMA POTTER. Fullerton, Calif.: Museum of North Orange County, 1975. 75 p. Illus. (part color). Bibliography, p. 73. Paper.

John E. Collins's commentary includes a brief biography of Lucy Lewis and a history of the Acoma pueblo. Her designs, methods, and materials are explained and some thirty of her pieces are illustrated.

Marriott, Alice. MARIA: THE POTTER OF SAN ILDEFONSO. The Civilization of the American Indian Series, vol. 27. Norman: University of Oklahoma Press, 1948. 294 p. Illus. Bibliography, pp. 291-94.

Based on a number of personal interviews with this well-known contemporary Indian potter, this informal biography of Maria Montoya Martinez chronicles her life and indirectly describes her work and the life of her pueblo.

Maryland Institute College of Art. CERAMICS 1973-1974. Baltimore: 1973. 21 p. 22 halftones. Paper.

The catalog of this student exhibition includes a list of exhibitors and large illustrations of seventeen pieces submitted.

Moore College of Art, Gallery. CLAY THINGS: EAST COAST INVITATIONAL. Philadelphia: 1974. 20 p. 9 halftones. Paper.

The catalog of this invitational exhibit contains a brief biography, one example, and a statement by each of the nine participating artists.

Natzler, Gertrud, and Natzler, Otto. CERAMICS: CATALOGUE OF THE COLLECTION OF MRS. LEONARD M. SPERRY AND A MONOGRAPH BY OTTO NATZLER.

See: Chapter 9: "Ceramic Collections" (p. 142).

_____. THE CERAMIC WORK OF GERTRUD AND OTTO NATZLER: A RETROSPECTIVE EXHIBITION. Los Angeles: Los Angeles County Museum of Art, 1966. 49 p. 27 illus. (8 color). Paper.

This exhibition catalog contains a number of very handsome illustrations of the work of this well-known couple.

_____. FORM AND FIRE: NATZLER CERAMICS. 1939-1972. Washington, D.C.: Smithsonian Institution Press for the Renwick Gallery of the National Collection of Fine Arts, 1973. 120 p. Illus. (some color). Bibliography, pp. 109-111. Paper.

Published for the retrospective exhibition of the Natzlers' work at the Renwick Gallery of the National Collection of Fine Arts, this large and handsome catalog illustrates many of the 150 pieces in the exhibition. Otto Natzler's annotations for several pieces are included as well as brief introductory essays by Natzler and Daniel Rhodes. Information on the dating of the Natzlers' pieces, lists of museum collections and exhibitions, and biographical notes complete the text.

New Mexico. University of. Maxwell Museum of Anthropology. 7 FAMILIES IN PUEBLO POTTERY. Albuquerque: 1974. 112 p. Illus. (part color). Glossary, pp. 108-11. Paper.

This exhibition catalog outlines the pottery-making tradition of seven Indian families in the American Southwest: the Chino and Lewis families (Acoma pueblo); the Gutierrez and Tafoya families (Santa Clara pueblo); the Gonzales and Martinez families (San Ildefonso pueblo); and the Nampeyo family (Hopi pueblo). In most cases, work by several members of two or three generations of potters is represented. The catalog is composed principally of illustrations and quotations from the various potters about their work and includes chronological tables of each of the families.

New York. State University of, at Geneseo. College of Art and Sciences. RAKU INVITATIONAL. Geneseo, N.Y.: 1970. 28 p. Halftones. Paper.

This catalog includes a biographical outline and photograph of each artist and reproduces an example of each artist's work. Thirteen artists, including Paul Soldner and Ken Ferguson, are represented.

New York. State University of, at New Paltz. College Art Gallery. CERAMICS:

1972, INTERCOLLEGIATE EXHIBITION. New Paltz, N.Y.: 1972. 46 p. Halftones. Paper.

Illustrations of most of the sixty-three objects in this student exhibition and biographical data on each of the artists are included in this catalog.

New York City. Museum of Contemporary Crafts. CLAYWORKS: 20 AMERICANS. New York: 1971. 36 p. 25 halftones. Paper.

Not intended as a survey of ceramics today, this exhibition shows works by a group of artists engaged in commentary on present-day society. The catalog illustrates at least one work by each artist, lists the ninety-seven pieces exhibited, and supplies twenty biographies, and a statement by each artist on his or her work.

Oakland Museum. CALIFORNIA CERAMICS AND GLASS 1974. Oakland, Calif.: 1974. Unpaged. Halftones. Paper.

This exhibition catalog includes brief biographies of each of the 271 competitively chosen artists in the exhibition and illustrates twenty-three of the award-winning entries.

Pierson, Conway. CONWAY PIERSON. Santa Barbara: University of California at Santa Barbara, Art Gallery, 1967. 61 p. Illus. (4 color). Paper.

This brief catalog illustrates an exhibition of Pierson's recent work in clay and bronze.

Poor, Henry Varnum. A BOOK OF POTTERY: FROM MUD TO IMMORTALITY. Englewood Cliffs, N.J.: Prentice-Hall, 1958. 192 p. 50 illus. (5 color, 45 halftone), 61 figs. Bibliography, p. 187. Index, pp. 189-92.

This book is a very personal explanation of the author's own philosophy and approach to the ceramic process and includes numerous drawings by the author and photographs of his work. Chapters are devoted to information on making pots, decoration, sculpture, glazing, and kilns.

Rippon, Ruth. RIPPON. Sacramento: E.B. Crocker Art Gallery, 1971. 62 p. Illus. (incl. 11 color). Paper.

This catalog of a retrospective exhibition of work by Ruth Rippon includes a number of her drawings and her ceramics of 1953-71.

San Francisco Potters, Association of; and Marin. College of. CLAY/GLASS '75. Kentfield, Calif.: 1975. 24 p. 24 halftones. Paper.

This catalog describes each of the fifty-seven pieces exhibited, including sixteen by invited artists, and illustrates several award-winning entries as well as each of the invited entries.

Scripps College. NINETEENTH ANNUAL INVITATIONAL CERAMICS EXHIBITION. [Claremont], Calif.: 1963. 34 loose pls. Paper.

Each plate of this catalog illustrates one object by each of the participating artists.

Smithsonian Institution. CERAMIC ARTS U.S.A. 1966. Washington, D.C.: 1966. 28 p. Halftones. Paper.

This catalog illustrates the one hundred pieces selected for this exhibition sponsored by the International Minerals and Chemical Corporation. A portfolio of illustrations in the center of the catalog indicates the thirty-two objects selected for exhibition only at the Smithsonian Institution.

Syracuse Museum of Fine Arts, American Ceramic Society. CONTEMPORARY AMERICAN CERAMICS. Syracuse, N.Y.: 1937. 18 p. 18 illus. (1 color). Paper.

This brief catalog lists the 135 objects assembled by the Syracuse Museum in 1937 for exhibition in Denmark, Sweden, and Finland. Some fifty living artists are represented, including Henry Varnum Poor, Maria Martinez, and Harold Riegger. While the illustrations do not do justice to the few pieces shown, they probably convey some sense of ceramic taste in the 1930s.

Wildenhain, Frans. FRANS WILDENHAIN RETROSPECTIVE. Binghamton, N.Y.: University Art Gallery, 1974. 44 p. 27 halftones. Paper.

Covering the period from 1933 to the present, this catalog includes a list of the 155 pieces in the exhibition, biographical information on the artist, and brief commentaries on his work by six artists and museum directors. Some two dozen pieces are attractively illustrated.

Wildenhain, Marguerite. THE INVISIBLE CORE: A POTTER'S LIFE AND THOUGHTS. Palo Alto, Calif.: Pacific Books, 1973. 207 p. 52 illus.

This book is both an autobiography, describing the author's experiences, including her years at the Bauhaus, and a series of essays explaining her philosophy and attitudes toward nature, art, education, and integrity.

_____. POTTERY OF MARGUERITE WILDENHAIN: A SELECTION OF HER RECENT WORK. Raleigh: North Carolina Museum of Art, 1968. 24 p. 18 illus. Paper.

This exhibition catalog includes a brief biography of the potter and several large illustrations of pieces in the exhibition.

Wisconsin. University of. Elvehjem Art Center. THE LARGE AND SMALL OF IT. Madison: 1971. 13 p. 4 halftones. Paper.

This very small catalog lists the sixty-two pieces exhibited by D.M. Reitz and Bruce Breckenridge and includes conversations with the artists about their work.

INTERNATIONAL CERAMICS

Alberta Potters Association. CERAMIC ART OF THE WORLD 1973. Calgary: 1973. The International Ceramic Exhibition and Conference. 120 p. Illus. (part color). Paper.

This large and handsomely illustrated catalog consists almost entirely of illustrations of the 250 ceramic pieces selected for this exhibition. A large number of countries are represented and separate sections are devoted to the prize-winning entries. Brief introductory remarks and the jury's report are also included.

American Federation of Arts. INTERNATIONAL EXHIBITION OF CERAMIC ART. Washington, D.C.: 1928-29. Unpaged. 31 halftones.

This catalog is introduced by Charles R. Richards and identifies each of the 404 ceramic pieces in the exhibition. Artists from Austria, Czechoslovakia, Denmark, England, France, Germany, Holland, Sweden, and the United States are represented.

Artigas, J. Llorens, and Matheos, J. Corredor. SPANISH FOLK CERAMICS: ACTUAL SPANISH FOLK CERAMICS OF TODAY. New York: Watson-Guptill Publications, 1974. 235 p. 36 color pls. 194 halftones, figs., map. Index, pp. 229-31. Bibliography, p. 235.

Extensively illustrated with large color and halftone photographs of potters, studios, villages, and wares, this handsome text introduces the reader to the techniques of Spanish folk ceramics, describing in separate chapters the plain and decorated wares of some thirteen pottery areas in Spain and the Balearic and Canary Islands. The authors informally describe the state of pottery production in hundreds of Spanish towns observed in their travels through the countryside and include a map of the pottery areas and a glossary of traditional vessel shapes.

Artner, Tivadar. MODERN HUNGARIAN CERAMICS. Budapest: Art Foundation Publishing House, 1974. 26 p. 19 color illus. Paper.

Published in six languages, this volume, with a preface by Artner, appears to be an exhibition catalog showing the work of several contemporary Hungarian artists. A brief identification accompanies each of the illustrations.

Australian Gallery Directors' Conference. AUSTRALIAN CERAMICS. N.p.: 1974. 48 p. Halftones. Bibliography. Paper.

This catalog identifies each of the 130 objects exhibited and provides a brief biographical outline and photograph of each of the forty-six participating artists. One example of each artist's work is also illustrated.

*Beard, Geoffrey. MODERN CERAMICS. Studio Vista/Dutton Picturebacks, David Herbert, general ed. New York: E.P. Dutton & Co.; London: Studio Vista, 1969. 167 p. 120 illus. (4 color). Glossary, p. 161. Bibliography, p. 163. Index, pp. 165-67. Paper.

This small picturebook shows numerous examples of the directions and developments in ceramics during the past sixty years. Work from California, Japan, England, Germany, and the Scandinavian countries is included.

Blattler, Hugo. CERAMICS IN ITALY. Arti Praticae Series. Rome: Aristide Palombi Editore, 1958. 206 p. Halftones.

Printed in four languages, this volume is composed almost entirely of illustrations of contemporary Italian ceramics. Also included are lists of the artists illustrated, museums with ceramic collections, ceramic competitions, and magazines.

Bryk, Rut. RUT BRYK. Milan: Lisa Licitra Ponti, 1970. 10 p. Illus. (8 color). Paper.

Beginning with a brief introductory essay, this catalog includes a biography of the artist, a list of her works in the exhibition, and several large illustrations of her wall sculptures.

Bunt, Cyril G.E. BRITISH POTTERS AND POTTERY TODAY. Leigh-On-Sea, Eng.: F. Lewis, Publishers, 1956. 78 p. 48 pls. (8 color). 4 appendixes, pp. 63-78.

In addition to a very brief history of English pottery during the past two centuries and four appendixes of potters working in 1710, 1787, 1802, and 1954, this volume presents a cross section of those companies manufacturing china in England today. A brief outline of each company is presented and the reproductions all show contemporary table services or figurines.

Cardozo, Sidney B. ROSANJIN: TWENTIETH CENTURY MASTER POTTER OF JAPAN. New York: Japan Society, 1972. 100 p. 75 illus. (6 color). Bibliography, p. 20. Paper.

This exhibition catalog gives a short biography of this remarkable potter and illustrates a number of examples of his versatile craftsmanship.

Casson, Michael. POTTERY IN BRITAIN TODAY. Concepts in Art series. London: Alec Tiranti, 1967. 22 p. 184 pls. (1 color).

This volume documents the revival of nonindustrial pottery in contemporary Britain. The book's three principal subject divisions are hand building, wheel-made pottery, and workshop production (making domestic ware in multiple). The work of approximately seventy-one potters is represented.

Digby, George Wingfield. THE WORK OF THE MODERN POTTER IN ENGLAND. London: John Murray, 1952. 110 p. 64 pls. Bibliography, pp. 92-94. Glossary, pp. 95-97. Index, pp. 108-10.

Written to acquaint more people with handmade English pottery, this book briefly explains the ceramic process and examines the points of view and standards which the artists who create this pottery bring to their work. Included are short biographies of some twenty English potters, including Bernard and David Leach, Michael Cardew, Lucie Rie, and Hans Coper.

Faenza, Italy. XXIII CONCORSO INTERNAZIONALE DELLA CERAMICA D'ARTE. Faenza: 1965. lxxvii, 275 p. 275 halftones. Paper.

This is the catalog of the Twenty-third International Competition of Ceramic Arts held at Faenza, Italy. The competition was begun in 1938 and has been conducted annually since then. Published in several languages (including English), the catalog contains brief introductions to the competition and many illustrations of works submitted by artists from some thirty nations. Having seen that catalogs were also published for the competitions of 1953, 1956, 1959, and 1964 through 1975, I assume that a similar catalog was issued for each of the competitions.

Forsyth, Gordon. TWENTIETH CENTURY CERAMICS. London: Studio; New York: Studio Publications, 1936. 128 p. Illus. (8 color). Index, pp. 124-28.

Written by the principal of the Stoke-on-Trent Art School, this international survey gives an interesting perspective on the changes in style, taste, and design in the twentieth century. While many of the pieces are still quite handsome, others have not withstood the passage of time quite so well.

Foster, George M. CONTEMPORARY POTTERY TECHNIQUES IN SOUTHERN AND CENTRAL MEXICO. Middle American Research Institute, no. 22. New Orleans: Tulane University, 1955. 36 p. 12 pls., figs., map. Bibliography, p. 33. Index, pp. 34-35. Paper.

The first part of this brief and very readable study examines the various techniques used in the forming and decorating of contemporary Mexican pottery. Part 2 discusses the different pottery-making centers in Mexico, state by state. According to the author, his study can only be viewed as a very introductory and incomplete guide because of the vast number of pottery centers in Mexico.

Holland, W. Fishley. FIFTY YEARS A POTTER. Pendley Manor, Tring, Hertsfordshire, Eng.: POTTERY QUARTERLY, 1958. 105 p. 11 illus.

This is an entertaining autobiography by a fifth-generation English "peasant" potter. Informally combining his personal history and philosophy, the author touches upon his boyhood years and then recounts some fifty years of pottery making, beginning in 1900.

Japan Pottery Design Center. 1973 INTERNATIONAL POTTERY DESIGN COMPETITION. Nagoya, Jap.: 1973. Unpaged. Illus. (part color). Paper.

Printed for the most part in English, this catalog includes information on the artists, judges, participants, and winners of this 1973 competition. The fifty or so prizewinning and honorably mentioned pieces are illustrated.

Kusube, Yaichi. THE CERAMIC ART OF YAICHI KUSUBE. Tokyo: Chuo-Koron Bijutsu Shuppan, 1968. 43 p. 100 pls. (28 color).

Written in both Japanese and English, this oversize volume includes a brief biography of the artist and a selection of large illustrations. The remainder of the text consists of captions identifying the numerous illustrations representing this craftsman's varied work.

Leach, Bernard. HAMADA: POTTER. Tokyo and New York: Kodansha International, 1975; distributed in the United States by Harper & Row, Publishers, New York. 306 p. Illus. Index, pp. 303-5.

Oversize and handsomely printed, this biography of Hamada is organized not as a traditional history but as a dialogue between Hamada and Leach from some three months of taped interviews. This technique has resulted in Hamada's telling his own life story, including reminiscences of his years working with Leach, his travels, various pottery ideas, and even his chair collection. An album of informal photos, a sketchbook of his designs, and a portfolio of his pots incorporating Hamada's brief comments are located in a chapter at the end of the volume, which aims to give insight into Hamada's thinking and way of life rather than such technical information as glaze recipes, etc.

_____. A POTTER IN JAPAN: 1952-1954. New York: Transatlantic Arts, 1967. 246 p. Illus. Glossary, pp. 239-40. Index, pp. 241-46.

This very enjoyable book is Leach's informal personal diary of his 1953-54 visit to Japan, which he made after an absence of many years. The book begins with a chapter on the first international conference of potters and weavers, held in England in 1952, which Leach helped organize.

______. THE POTTER'S CHALLENGE. Edited by David Outerbridge. A Sunrise Book. New York: E.P. Dutton & Co., 1975. 156 p. Illus. Museum directory, pp. 139-51. Bibliography, pp. 153-56.

The chapters of the first part of this two-part book are the result of several interviews between Leach and Outerbridge and deal with a variety of Leach's ceramic ideas. Part 2, entitled "Exemplary Pots," is taken from material published by Leach in A POTTER'S PORTFOLIO (p. 12) and contains Leach's informal comments on each of the pieces illustrated. David MacKenzie's briefly annotated directory of U.S. and world museums with ceramic collections is also included.

______. A POTTER'S WORK. [London]: Evelyn, Adams & Mackay, 1967; distributed in the United States by Kodansha International, Palo Alto, Calif. 150 p. 80 pls. (12 color), drwgs.

Beginning with an introduction and a biography of Leach by J.P. Hodin, this volume comprises a photographic collection of Leach's work, selected by the artist and supplemented with thirty pages of his notebook drawings.

*Lewenstein, Eileen, and Cooper, Emmanuel, eds. NEW CERAMICS. New York: Van Nostrand Reinhold Co., 1974. 224 p. 16 color pls., 227 halftones. Bibliography, p. 220. Index, pp. 221-24.

Composed almost entirely of large and handsome illustrations, this volume comprehensively surveys current trends in studio pottery throughout the world. Ceramic objects from over twenty countries are represented and each country's work is introduced in an essay written by a native artist or an individual involved in the arts.

Lister, Raymond. DECORATED PORCELAINS OF SIMON LISSIM. Cambridge, Eng.: Golden Head Press, 1955. 32 p. 13 pls., marks. Bibliography, pp. 21-26.

Relating Lissim's design to his origins in Russia and his early artistic training, Lister discusses the forms and imagery with which Lissim decorates his brilliantly painted porcelain pieces.

Litto, Gertrude. SOUTH AMERICAN FOLK POTTERY. New York: Watson-Guptill Publications, 1976. 224 p. 30 color pls. 275 halftones and figs., maps. Bibliography, p. 221. Glossary/Index, pp. 221-24.

Centered around a journey taken by the Litto family through the west coast of South America, this volume examines the continuing pottery tradition in various parts of Peru, Bolivia, Chile, Ecuador, Columbia, and Venezuela. Litto treats each town or pottery locale in some detail, describing various families, the materials and equipment they use, marketing practices, and both old and new influences on the different styles of ware produced. In addition to numerous illustrations of different pottery types, there are many halftones and figure illustrations of the tools, equipment, and forming techniques used in the different areas.

THE FRED AND MARY MARER COLLECTION.

See: Chapter 9: "Ceramic Collections" (p. 141).

Octagon Center for the Arts. TWELVE DUTCH POTTERS. Ames, Iowa: 1971. 31 p. Illus. Paper.

This small catalog includes Bernadine de Neeve's introduction to the exhibition and a two-page resume of each artist with one illustration of his or her work.

Pataky-Brestyanszky, Ilona. MODERN HUNGARIAN CERAMICS. Translated by Lili Halapy. Budapest: Corvina Press [1961]. 46 p. 48 pls. (8 color). Bibliography, p. 44.

This short volume examines the work of several Hungarian artists from the 1930s to 1960, placing special emphasis on the work of Istvan Gavor, Geza Gorka, and Margit Kovacs.

Peterson, Susan. SHOJI HAMADA: A POTTER'S WAY AND WORK. Tokyo and New York: Kodansha International, 1974; distributed in the U.S. by Harper & Row, Publishers, New York. 239 p. 271 illus. (18 color). Appendix, pp. 213-33. Glossary-index, pp. 234-39.

Written by a longtime friend of Hamada, who spent four months at his compound in Mashiko, this warm portrait loosely follows the ceramic cycle, beginning with the workshop and working of the clay, then the glazing and firing, and, finally, the evaluation of the finished pots. Other sections of the text describe Hamada's museum and his attendance at a Tokyo exhibition. Throughout the text Hamada comments on his work and expresses his feelings on a wide variety of ideas.

Ramie, Suzanne, and Ramie, Georges. CERAMIQUES DE PICASSO. Geneva: Editions d'Art Albert Skira, 1948. 20 p. 18 color pls. Portfolio.

This portfolio contains a thirteen-page introduction in French to the ceramic art of Picasso and eighteen mounted illustrations of his painted (glazed) plates. The pieces include some of his better-known themes, including the owl, dove, and bullfighter.

Rose, Muriel. ARTIST POTTERS IN ENGLAND. Faber Monographs on Pottery and Porcelain. London: Faber and Faber, 1970. 64 p. 123 pls. (6 color). 146 illus. Marks, pp. 54-55. Bibliography, pp. 56-61. Index, pp. 62-64.

Beginning with chapters describing the work of several early artist-potters in England and France during the late nineteenth and early twentieth centuries, Rose continues with brief histories of about twenty of the most famous contemporary English potters, including Bernard Leach, Michael Cardew, Shoji Hamada, Lucie Rie, and Hans Coper. Although the work of each artist is illustrated, most of the illustrations reproduce work by those who are better known.

Stitt, Irene. JAPANESE CERAMICS OF THE LAST 100 YEARS.

See: Chapter 4: "Eastern Ceramics" (p. 42).

Syracuse University School of Art. CONTEMPORARY JAPANESE CERAMICS. Syracuse, N.Y., [196?]. Unpaged. Illus. Paper.

This catalog of an exhibition organized by Professor Ruth Randall of ceramics exhibited for the first time as part of the university's permanent collection includes an introductory essay on the tradition and challenge of contemporary Japanese ceramics by Charles R. Dibble.

Tennessee Arts Commission. INTERNATIONAL CERAMICS SYMPOSIUM: MEMPHIS '73. Memphis: 1973. International Academy of Ceramics. 56 p. Halftones. Paper.

Included in this catalog are a photograph of each of the twenty-three international participants, one large illustration of each artist's work, and each artist's brief statement or description of his or her work.

Torbert, Meg, ed. DESIGN QUARTERLY. Nos. 42-43, 1958. Halftones.

This double issue is composed entirely of biographical sketches of some eighty contemporary potters and of numerous illustrations of their work. Many of the sketches include comments by the artists concerning their personal convictions about the art of ceramics.

Vallauris, France. BIENNAL INTERNATIONALE DE LA CERAMIQUE D'ART. Vallauris: 1968. Illus. Paper.

Printed in both French and English, this catalog illustrates ceramic works submitted to this international exhibition by artists from some twenty countries.

Victoria and Albert Museum. INTERNATIONAL CERAMICS 1972. London: 1972. 122 p. Illus. Paper.

Arranged alphabetically by country, this catalog includes many small illustrations of pieces submitted by craftsmen from thirty-nine countries.

Virginia. University of. Mary Washington College. THE POTTERY OF TERUO HARA. Fredericksburg, Va.: DuPont Galleries, 1969. 12 p. 9 halftones. Paper.

This exhibition catalog includes a two-page summary by Pauline G. King of Hara's life and work. Teruo Hara came to the United States from Japan in 1958 and, at the time of this exhibit, had been a member of the faculty at the Corcoran School of Art in Washington, D.C., for five years.

Washington, D.C. Kiln Club. TENTH INTERNATIONAL EXHIBITION OF CERAMIC ART. Washington, D.C.: Smithsonian Institution, Museum of History and Technology, 1965. 8 p. Illus. Paper.

This is the catalog of the tenth biennial exhibition sponsored by the Kiln Club of Washington, D.C. About twenty award-winning pieces from the exhibition are illustrated.

Widman, Dag, and Klyvare, Berndt. STIG LINDBERG--SWEDISH ARTIST AND DESIGNER. Translated by P.A. Burke. Stockholm: Raben and Sjogren, 1962. Unpaged. Illus. (part color).

This volume examines the wide range of Lindberg's work as an industrial designer, from the unique pieces produced at the State School of Arts, Crafts and Design to the anonymous, mass-produced objects from the Gustavsbergs Fabriker, the largest pottery works in Sweden.

Willcox, Donald J. NEW DESIGN IN CERAMICS. New York: Van Nostrand Reinhold Co., 1970. 119 p. 229 illus. (19 color).

Most of this volume consists of numerous photos attesting to the originality of contemporary Scandinavian ceramics; the brief text introduces the reader to the potter's role in Scandinavia and to the basic clay-working process.

Yanagi, Soetzu. THE UNKNOWN CRAFTSMAN: A JAPANESE INSIGHT INTO BEAUTY.

See: Chapter 4: "Eastern Ceramics" (p. 42).

Chapter 9
CERAMIC COLLECTIONS

Publications without a named individual or individuals as author have been cataloged under the name of the museum owning the collection or the name of the individual collector, whichever seemed more suitable.

Barber, Edwin Atlee. CATALOGUE OF AMERICAN POTTERIES AND PORCELAINS. Philadelphia: Pennsylvania Museum and School of Industrial Art, 1893. 43 p. Illus. Paper.

> E.A. Barber's collection of American ceramics, which had been formed for the purpose of illustrating his HISTORY OF THE POTTERY AND PORCELAIN OF THE UNITED STATES, was sold to the Pennsylvania Museum. This volume is a preliminary catalog of that collection, identifying both artistic and technical examples of American pottery. Brief explanations of the different types of ware are included.

______. HISPANO-MORESQUE POTTERY. Publications of the Hispanic Society of America, no. 94. New York: Hispanic Society of America, 1915. 278 p. 88 halftones.

> Beginning with an introduction to the origins and techniques of these wares, Barber catalogs in some detail the 132 vases, plates, tiles, etc., in the society's collection.

______. SPANISH MAJOLICA IN THE COLLECTION OF THE HISPANIC SOCIETY OF AMERICA. New York: Hispanic Society of America, 1915. 150 p. 152 halftones. Paper.

> Including a brief introduction by Barber, this volume is an annotated catalog of the Hispanic Society's collection of Spanish majolica.

______. SPANISH PORCELAINS AND TERRA COTTAS IN THE COLLECTION OF THE HISPANIC SOCIETY OF AMERICA. Publications of the Hispanic Society of America, no. 93. New York: Hispanic Society of America, 1915. 42 p. 11 pls. (1 color).

Beginning with a brief history of the Bien Retiro factory near Madrid, Barber catalogs the twenty-two pieces of porcelain and terracotta in the society's collection from Bien Retiro and Capo di Monte.

Beazley, J[ohn].D[avidson]. ATTIC RED-FIGURED VASES IN AMERICAN MUSEUMS. Cambridge, Mass.: Harvard University Press, 1918. x, 236 p. 118 halftones and figs. Appendix, pp. 193-236.

Considering vases from the end of the sixth century B.C. to the beginning of the fourth century A.D., Beazley attempts to assign a painter to each of the red-figured vases in American museum collections. Each assigned American vase identifies the location of several additional vases by the same painter that are not in American collections.

_____. GREEK VASES IN POLAND. Oxford: Clarendon Press, 1928. 87 p. 32 halftones. Index, pp. 82-87.

This extensively footnoted volume examines pieces from three principal Polish collections: that of Prince Czartoryski, the Czartoryski Museum at Cracow, and the teaching collection at the University of Cracow. Black-figured, Archaic red-figured, freestyle red-figured, and later red-figured vases make up the four divisions of Beazley's text.

Boardman, John, and Pope, Maurice. GREEK VASES IN CAPE TOWN. Guide, no. 6. Cape Town: South African Museum, 1961. 20 p. 31 halftones in 16 pls. Bibliography, p. 20. Paper.

This museum catalog illustrates and describes in some detail approximately twenty red-figured and black-figured vases in the museum's collection. Twelve other vases are briefly identified but not illustrated.

British Museum. A GUIDE TO THE POTTERY AND PORCELAIN OF THE FAR EAST IN THE DEPARTMENT OF CERAMICS AND ETHNOLOGY. London: 1924. 168 p. 14 pls., 230 illus. Index, pp. 165-68. Paper.

Although dated in a number of ways, this guide does survey briefly the history of oriental ceramics and conveys some sense of the scope of the museum's holdings.

Brown, A[nn].C[ynthia]. CATALOGUE OF ITALIAN TERRA-SIGILATA IN THE ASHMOLEAN MUSEUM. Oxford: Clarendon Press, 1968. xix, 39 p. 25 pls. figs. Bibliography, pp. xiii-xiv.

This catalog of Arretine wares groups 114 examples (mostly fragments) around seven known artists or workshops. Each of the workshops is briefly identified and the fragments are individually identified, described, and illustrated.

Brundage Collection, Avery. AVERY BRUNDAGE COLLECTION: CHINESE CERAMICS. San Francisco: deYoung Museum Society, 1967. 166 p. 75 pls. (16 color). Bibliography, pp. 163-66. Paper.

> This catalog of part of the Brundage collection at the M.H. de-Young Museum annotates its numerous illustrations and includes a brief introduction by the museum's director, Rene-Yvon Lefebvre d'Argence.

Caskey, L[acey].D[avis]. ATTIC VASE PAINTINGS IN THE MUSEUM OF FINE ARTS, BOSTON, part 1. Boston: Museum of Fine Arts, 1931. 57 p. 39 figs., 10 halftones in 4 pls. Paper.

> Written in cooperation with J.D. Beazley, this first of a three-part series covering the museum's collection of Attic vases identifies and describes some sixty-five examples.

Caskey, L[acey].D[avis]., and Beazley, J[ohn].D[avidson]. ATTIC VASE PAINTINGS IN THE MUSEUM OF FINE ARTS, BOSTON, part 2. Boston: Museum of Fine Arts, 1954. 103 p. Figs., 29 halftones in 12 pls. Paper.

> Completed by Beazley after Caskey's death in 1944, part 2 begins with number 66 and covers forty-eight more items in the museum's collection of painted Attic vases.

_____. ATTIC VASE PAINTINGS IN THE MUSEUM OF FINE ARTS, BOSTON, part 3. Boston: Museum of Fine Arts, 1963. 93 p. Figs., 38 halftones in 12 pls. Paper.

> Including some sixty-two additional examples, this is the concluding part of the museum's study of its Attic vase collection. (For parts 1 and 2, see entries under Caskey, L.D. (above).

Cheng, Te-K'un. AN INTRODUCTION TO SZECHWAN POTTERY. Guidebook Series, no. 2. Chengtu: West China Union University Museum, 1945. 21 p. Paper.

> This catalog briefly describes the museum's holdings that were on display in 1945, generally dividing them into eight stages of ceramic development.

Clement, Arthur W. NOTES ON AMERICAN CERAMICS: 1607-1943. Brooklyn, N.Y.: Brooklyn Museum, 1944. 63 p. 56 halftones. Bibliography, pp. 35-36. Paper.

> Designed as a small handbook for the Brooklyn Museum's collection of American earthenware and porcelain, this pamphlet examines the following principal types of American ceramics: redware, stoneware, moulded wares, utility wares, decorative wares, porcelains, and tiles.

Cook, Brian F. INSCRIBED HYDRA VASES IN THE METROPOLITAN MUSEUM OF ART. Paper no. 12. New York: Metropolitan Museum of Art, 1966. 36 p. 17 pls. 2 appendixes, pp. 33-34. 3 concordances, p. 35. 2 indexes, p. 36.

Including an introduction to the history of the vases, this catalog identifies twenty-seven hydriae in the museum's collections, including its date, description, inscription in both Greek and English, and bibliography. Each vase is illustrated at least once.

DeLaBeche, Sir Henry, and Reeks, Trenham. CATALOGUE OF SPECIMENS ILLUSTRATIVE OF THE COMPOSITION AND MANUFACTURE OF BRITISH POTTERY AND PORCELAIN. London: Museum of Practical Geology, 1855. xxiii, 179 p. 150 figs.

This very old catalog describes the use of various local mineral substances in the manufacture of ceramic wares from the time of the Roman occupation to the mid-nineteenth century. Several chapters describe the general characteristics of clays and glazes while some earlier foreign wares, such as Greek vases and majolica, are included to show the development and use of enameled colors.

deRicci, Seymour. A CATALOGUE OF EARLY ITALIAN MAJOLICA IN THE COLLECTION OF MORTIMER L. SCHIFF. New York: privately printed, 1927. Halftones.

Dealing with Italian majolica, 1400-1530, this large catalog describes and illustrates 111 pieces in the Schiff collection. Many of the objects are illustrated in two full-page plates to show front and back (or side) views, and the description accompanying each object includes details on date, place of manufacture, and other matters of provenance.

Dibble, Charles Rynder. THE JOHN R. FOX COLLECTION OF KOREAN CERAMICS AT SYRACUSE UNIVERSITY. Syracuse, N.Y.: Syracuse University School of Art, 1965. 108 p. Illus. Glossary, pp. 101-5. Bibliography, pp. 106-7. Paper.

This museum catalog illustrates Syracuse University's Fox collection of Korean ceramics from the Silla through Yi dynasties. Dibble's seventeen-page history of Korean ceramic wares is included and each piece illustrated is clearly described.

Duckworth Collection, Mr. and Mrs. Harold. ENGLISH AND AMERICAN CERAMICS OF THE EIGHTEENTH AND NINETEENTH CENTURIES: A SELECTION FROM THE COLLECTION OF MR. AND MRS. HAROLD G. DUCKWORTH. Toledo, Ohio: Toledo Museum of Art, 1968. 53 p. 12 halftones. Bibliography, pp. 52-53. Paper.

Although quite small and containing a limited number of illustrations, this catalog does include Assistant Curator John Webster Keefe's eighteen-page introduction to the various wares shown in the exhibition at the Toledo Museum. The catalog is organized by ceramic type and describes some 220 ceramic pieces.

Eriksen, Svend. SEVRES PORCELAIN: THE JAMES A. DEROTHSCHILD COLLECTION AT WADDESTON MANOR. London: National Trust for Places of Historic Interest or Natural Beauty, 1968. 339 p. Illus. (part color).

While including a brief account of the early history of Sevres and descriptions of throwers, gilders, painters, marks, and colors, this volume largely illustrates and describes in detail some 150 pieces of Sevres given to the British government by James de-Rothschild.

Fairbanks, Arthur. CATALOGUE OF GREEK AND ETRUSCAN VASES: EARLY VASES, PRECEDING ATHENIAN BLACK-FIGURED WARE. Cambridge, Mass.: Harvard University Press for the Museum of Fine Arts, Boston, 1923. 235 p. 100 pls.

This catalog identifies and describes some seven hundred vases and a number of fragments in the Museum of Fine Arts, Boston's collection. Examples of Minoan, Mycenaean, Cypriot, Thessalian, Geometric, early black-figure, Italian bucchero, and Italian Geometric vases are included and many are illustrated.

Fehervari, Geza. ISLAMIC POTTERY. London: Faber and Faber, 1973. 191 p. 330 illus. in 138 pls. (18 color). Map. Bibliography, pp. 181-83. Index, pp. 187-91.

This comprehensive study of Sir Alan Barlow's now-dispersed collection of Islamic pottery includes and illustrates every piece in the original collection. Each period style is introduced and all appropriate pieces are described.

Fourcade, François. ART TREASURES OF THE PEKING MUSEUM. Translated by Norbert Guterman. New York: Harry Abrams, Publishers, 1965. 177 p. 87 color pls.

Although principally devoted to painting, this volume does contain excellent illustrations of thirty-seven fine examples of Chinese ceramics ranging from the Neolithic era to the eighteenth century.

Freer Gallery of Art. MEDIEVAL NEAR EASTERN CERAMICS: IN THE FREER GALLERY OF ART. Smithsonian Publication, no. 4420. Washington, D.C.: Smithsonian Institution, 1960. 38 p. 40 halftones. Paper.

This short picture book of examples from the Freer Gallery collection includes a five-page introduction by Dr. Richard Ettinghausen, curator of Near Eastern art at the Freer.

Gerry Collection, Capt. and Mrs. Roger. JAPANESE CERAMICS FROM THE COLLECTION OF CAPTAIN AND MRS. ROGER GERRY. New York: Brooklyn Museum, 1961. Unpaged. 24 halftones. Bibliography. Paper.

> Beginning with a seven-page introduction by Captain Gerry, this exhibition catalog includes the Gerrys' ceramic examples ranging from prehistoric times to the seventeenth century.

Gordon, G[eorge].B[yron]., ed. EXAMPLES OF MAYA POTTERY IN THE MUSEUM AND OTHER COLLECTIONS. Philadelphia: University of Pennsylvania Museum, 1925. 66 color drwgs. Oversize portfolio, 15" x 21".

> This very large portfolio of colored illustrations shows the decorative figures and patterns on Maya pottery. The plates are identified but not accompanied by text.

Hackenbroch, Yvonne. CHELSEA AND OTHER ENGLISH POTTERY, PORCELAIN AND ENAMEL IN THE IRWIN UNTERMEYER COLLECTION. Cambridge, Mass.: Harvard University Press for the Metropolitan Museum of Art, 1957. xxx, 286 p. 346 illus. (part color). Bibliography, pp. 275-83. Index, pp. 285-86.

> This second volume of the catalog of the Untermeyer collection contains a brief introduction by Hackenbroch, and illustrates and identifies in some detail the history of each piece of English pottery in the Untermeyer collection.

_____. MEISSEN AND OTHER CONTINENTAL PORCELAIN, FAIENCE AND ENAMEL IN THE IRWIN UNTERMEYER COLLECTION. Cambridge, Mass.: Harvard University Press for the Metropolitan Museum of Art, 1956. 264 p. 240 illus. (part color). Bibliography, pp. 255-62. Index, pp. 263-64.

> This is the first of several illustrated volumes cataloging the Untermeyer collection; it includes Hackenbroch's brief introduction and an extensive bibliography. Each of the pieces in this category of the collection is illustrated and described in some detail.

Henderson, Gregory. KOREAN CERAMICS: AN ARTS VARIETY. Ohio State University, Division of Art Galleries, 1969. 60 p. 143 illus., map. Bibliography, p. 18. Paper.

> This catalog of the Henderson collection exhibited at Ohio State University includes an introduction by Mr. Henderson on the history and characteristics of Korean pottery. The pieces in the collection date from the first to the nineteenth centuries and each is illustrated and carefully described in the catalog.

Hobson, R[obert].L[ockhart]. CATALOGUE OF THE COLLECTION OF ENGLISH POTTERY IN THE DEPARTMENT OF BRITISH AND MEDIEVAL ANTIQUITIES AND ETHNOGRAPHY OF THE BRITISH MUSEUM. London: British Museum, 1903. xxiii, 310 p. 42 pls. (part color), 132 figs. Index, pp. 299-310.

Hobson briefly introduces each type of ware in the catalog which numbers and identifies every example of that type in the collection. Medieval tiles, sixteenth-century wares, and a wide variety of seventeenth- and eighteenth-century pottery types are represented in the collection.

_____. CATALOGUE OF THE LEONARD GOW COLLECTION OF CHINESE PORCELAIN. London: Leonard Gow, 1931. xxxix, 109 p. 85 pls. (part color). Bibliography, p. xxxix. Index, pp. 105-8.

This large and richly printed catalog (only 300 copies) identifies and describes the 421 pieces in the Gow collection and illustrates many of them. Brief essays on porcelain shapes, marks, and each of the various styles in the collection are also included.

_____. HANDBOOK OF THE POTTERY AND PORCELAIN OF THE FAR EAST. London: British Museum, 1937. 180 p. 260 halftones. Index, pp. 176-80.

More detailed than most such handbooks, this guide to the Far Eastern pottery and porcelain in the Department of Oriental Antiquities and Ethnography at the British Museum covers Chinese, Japanese, Korean, and Siamese wares. Several pages illustrate a number of Chinese and Japanese marks and such subjects as ceramic processes, provincial porcelain, and European influences are treated briefly.

Johnson, Franklin P. THE FARWELL COLLECTION. Monographs on Archaeology and Fine Arts, vol. 6. Cambridge, Mass.: Archaeological Institute of America, 1953. 76 p. 90 halftones in 20 pls.

This study examines the origins and styles of hand- and wheel-made pottery excavated from ancient graves by Captain Farwell in Italy during 1944-45. This collection contains perhaps the largest number of Daunian pottery pieces outside of Italy.

Lane, Arthur. A GUIDE TO THE COLLECTION OF TILES: VICTORIA AND ALBERT MUSEUM, DEPARTMENT OF CERAMICS. London: Board of Education, 1939. 75 p. 48 halftones. Subject and numerical indexes, pp. 66-75.

This guide contains Arthur Lane's brief history of tiles and identifies a number of examples from the extensive collections of the Victoria and Albert Museum.

_____. ISLAMIC POTTERY FROM THE NINTH TO THE FOURTEENTH CENTURIES IN THE COLLECTION OF SIR ELFRED HITCHCOCK. London: Faber and Faber, 1956. 36 p. 69 illus. in 40 pls. (6 color), map. Bibliography, p. 35.

Beginning with a brief introduction to the political and ceramic history of this Islamic period, Lane's catalog identifies and describes the more than seventy pieces in the Hitchcock collection. Almost all of the collection is illustrated.

LeCorbeiller, Clare. CHINA TRADE PORCELAIN: PATTERNS OF EXCHANGE. New York: Metropolitan Museum of Art, 1974. 134 p. Illus. (part color). References, p. 125. Index, pp. 127-34.

This volume examines some fifty-two pieces and groups of pieces added since 1955 to the Metropolitan's collection of China trade porcelain. Such an examination shows that these wares underwent style changes, had various European influences, and derived their painted decorations from several sources. All of these pieces are well illustrated and described in some detail in accompanying notes.

Leff Collection, Jay C. ANCIENT ART OF LATIN AMERICA: FROM THE COLLECTION OF JAY C. LEFF. New York: Brooklyn Museum, 1966. 139 p. Illus., maps, tables. Bibliography, p. 137. Paper.

Dealing almost exclusively with ceramic sculpture, this exhibition catalog contains a large selection of illustrations. Because these objects are from a private collection, they are often unusual and unique rather than the more typical objects generally chosen for museum collections.

Leith-Ross, Sylvia. NIGERIAN POTTERY: A CATALOGUE. Ibadan, Nigeria: University Press for the Department of Antiquities, [1970]. 200 p. Chiefly illus., folded map.

This catalog of pottery of the Jos Museum in Jos, Nigeria, is out of print and not available for a more detailed annotation. It has been included, however, because no comparable material appears available on this subject.

Lister, Robert H., and Lister, Florence C. THE EARL H. MORRIS MEMORIAL POTTERY COLLECTION: AN EXAMPLE OF TEN CENTURIES OF PREHISTORIC CERAMIC ART IN THE FOUR CORNERS COUNTRY OF SOUTHWESTERN UNITED STATES. Series in Anthropology, no. 16. Boulder: University of Colorado Press, 1969. ix, 94 p. 46 halftones, 1 color pl., 2 maps. Bibliography, p. 94. Paper.

Intended as part of a series of illustrated catalogs describing the collections of the University of Colorado, this well-illustrated text describes in some detail a representative sampling of pieces from the Earl Morris collection. The introduction to the collection and most of the descriptive material on each example were written by Morris himself.

McClellan, George B. A GUIDE TO THE McCLELLAN COLLECTION OF GERMAN AND AUSTRIAN PORCELAIN. New York: Privately printed, 1946. 36 p. 48 halftones in 24 pls. Paper.

> Printed to coincide with the first showing of the McClellan collection at the Metropolitan Museum of Art, to which the 287-piece collection was presented in 1941, this resume was prepared by McClellan himself before his death in 1940. In addition to its brief essays on the pleasures of collecting and "the great collections," the resume briefly describes the histories and products of several of the great factories, including Meissen, Vienna, Hochst, and Nymphenburg. While a number of pieces are illustrated, the entire collection is not cataloged here.

Manchester. City Art Gallery of. THE INCOMPARABLE ART: ENGLISH POTTERY FROM THE THOMAS GREG COLLECTION. Manchester, Eng.: 1969. 79 p. Halftones. Bibliography, p. 7. Paper.

> This exhibition catalog describes 222 pieces of English pottery from the Greg collection, which includes every type of English pottery from medieval times to the end of the eighteenth century. A large number of pieces are illustrated and the description of each piece is often quite detailed.

Marer Collection, The Fred and Mary. THE FRED AND MARY MARER COLLECTION. [Clairmont], Calif.: Scripps College, Lang Art Gallery, 1974. Unpaged. Illus. (part color). Paper.

> Beginning with two brief essays on Fred Marer and his collection by Jim Melchert and Paul Soldner, this handsome catalog identifies each of the 197 pieces exhibited. Organized by artist and including contemporary works from the United States, Europe, and Japan, the catalog illustrates approximately seventy pieces.

Miller, J. Jefferson. EIGHTEENTH CENTURY ENGLISH PORCELAIN: A BRIEF GUIDE TO THE COLLECTION IN THE MUSEUM OF HISTORY, AND TECHNOLOGY. Washington, D.C.: Smithsonian Institution Press, 1973. 95 p. 80 halftones. Bibliography, p. 95.

> This text is not intended as a definitive catalog of the collection but as an introduction to some of its more interesting pieces. Brief histories of each of the early English porcelain factories accompany the illustrations and a small bibliography for each factory is also included. The illustrations are large and generally of good clarity.

Monsen Collection, Prof. and Mrs. R. Joseph. A DECADE OF CERAMIC ART, 1962-1972: FROM THE COLLECTION OF PROFESSOR AND MRS. R. JOSEPH MONSEN. San Francisco: San Francisco Museum of Art, 1972. 60 p. 3 color pls., 53 halftones. Paper.

Introduced by Susan Foley, this exhibition catalog documents the growth of new expression in ten years of American ceramic art.

Morse, Edward S. CATALOGUE OF THE MORSE COLLECTION OF JAPANESE POTTERY. Cambridge, Mass.: Riverside Press, 1901. 384 p. 68 pls. 3 indexes, pp. 369-84.

This enormous catalog identifies the more than 5,000 pieces of the Morse collection in the Museum of Fine Arts, Boston. Several hundred marks are included and many of the pieces are photographed, some arranged in small groups, most in full display cases. Since publication, the Morse collection has become known as one of the most infamous examples of inaccurate attribution and not the complete survey of Japanese pottery that Morse intended. It contains much mediocre material and many nineteenth-century imitations.

National Palace Museum and National Central Museum. BLUE AND WHITE WARES OF THE MING DYNASTY. 5 vols. Hong Kong: Cafa Co., 1963. Chiefly illus. Color pls.

Printed in both Chinese and English, these oversize volumes are part of a continuing series publishing all the best porcelains in both museum collections. Except for their very brief introduction to Ming porcelains, these volumes are composed of a large color illustration and description of each piece. Considerable variation appears in the quality of the color of the plates.

Natzler, Gertrud, and Natzler, Otto. CERAMICS: CATALOGUE OF THE COLLECTION OF MRS. LEONARD M. SPERRY AND A MONOGRAPH BY OTTO NATZLER. Los Angeles: Los Angeles County Museum of Art, 1968. 81 p. Pls. (part color). Paper.

Intended as a complete catalog of the Sperry collection, this well-illustrated catalog is also a very personal document of the Natzlers' work. Separate essays by Otto Natzler describe the couple's early years in Vienna, Gertrud's ideas on form, ceramic appreciation, and the clay bodies and glazes used by the Natzlers. Each piece in the Sperry collection is identified, illustrated, and critically described by Otto Natzler.

New York City. Museum of Primitive Art. SCULPTURE FROM PERU: SELECTED FROM THE COLLECTION OF THE MUSEUM OF PRIMITIVE ART. New York: 1964. 31 p. 27 illus. Paper.

Of the twenty-seven sculptures in this small pamphlet, about half are clay works representing the styles of Peruvian art.

Orange, James. BIZEN-WARE: WITH A CATALOGUE OF THE CHATER COLLECTION. Yokohama and Shanghai: Kelly & Walsh, 1916. 115 p. 47 collotypes.

Dealing principally with ornamental and sculptured work rather than with tea ceremony utensils, this catalog illustrates and briefly describes some 162 pieces in the collection of Sir C.P. Chater. Also included are a brief history of the ware, lists of potters and signatures, and brief explanations of a number of Japanese folk legends relating to the ware.

Phillips, John Goldsmith. CHINA-TRADE PORCELAIN. Cambridge, Mass.: Harvard University Press for the Winfield Foundation and Metropolitan Museum of Art, 1956. xxi, 234 p. 17 color pls., 102 halftones. 57 figs. Appendix, pp. 215-21. Bibliography, pp. 222-24. References, pp. 225-27. Index, pp. 229-34. (Chamberlin, no. 1767).

Written by the Metropolitan's curator of Renaissance art, this extensively illustrated volume studied the Helena Woolworth McCann collection before its permanent distribution to twenty-six museums in the United States and one in Canada. The first part of the volume deals with Chinese porcelain centers and the history of trade porcelains; the second and larger part with the various styles and periods of the wares forming the McCann collection.

Pope, John Alexander. CHINESE PORCELAINS FROM THE ARDEBIL SHRINE. Washington, D.C.: Smithsonian Institution, Freer Gallery of Art, 1956. xv, 194 p. 142 pls., figs. Appendix, pp. 159-62. Bibliography, pp. 163-71. Index, pp. 173-94.

After introducing the reader to the origin and history of the Ardebil collection in Iran, this extensive and well-illustrated study examines in detail the periods of the various pieces making up the collection today. Since the blue-and-white pieces form the bulk of the collection, they are generally the focus of this study.

_____. FOURTEENTH-CENTURY BLUE-AND-WHITE: A GROUP OF CHINESE PORCELAINS IN THE TOPKAPU SARAYI MUZESI, ISTANBUL. Freer Gallery of Art Occasional Papers, vol. 2, no. 1. Washington: Smithsonian Institution, 1952. 85 p. 44 pls. 3 appendixes, pp. 66-73. Bibliography, pp. 74-78. Index, pp. 79-85. Paper.

By examining a number of their features, particularly their decoration, this paper identifies the David vases (in Sir Percival David's collection) and a group of some eighty pieces in the Topkapu Museum as a particular fourteenth-century blue-and-white style. About forty of these pieces are illustrated and accompanied by description.

Rackham, Bernard. A BOOK OF PORCELAIN: FINE EXAMPLES IN THE VICTORIA AND ALBERT MUSEUM. London: Adam & Charles Black, 1910. xv, 95 p. 28 color pls. Index, pp. 91-95.

Primarily this is a picture book of twenty-eight watercolor drawings done by William Gibb; Rackham's text simply introduces the reader to the period and style of each of the Chinese, Japanese, Italian, French, German, and English examples selected for illustration.

_____. CATALOGUE OF THE GLAISHER COLLECTION OF POTTERY AND PORCELAIN IN THE FITZWILLIAM MUSEUM, CAMBRIDGE. 2 vols. Cambridge: At the University Press, 1935. Vol. 1: xx, 431 p. 36 color pls. Bibliography, pp. xvii-xviii. Index, pp. 419-31. Vol. 2: 266 halftones.

These large and extensively illustrated volumes catalog the more than 3,000 ceramic pieces in the Glaisher collection. Information on marks, place, and time of acquisition, etc., is supplied with the description of each object and some one thousand pieces are illustrated in the second volume. English and European wares are the strengths of the collection while Japanese and Chinese pieces are hardly represented.

_____. CATALOGUE OF THE HERBERT ALLEN COLLECTION OF ENGLISH PORCELAIN. 2d ed. London: Victoria and Albert Museum, 1923. xvi, 172 p. 99 pls. Bibliography, pp. xi-xiii. Index, pp. 165-72.

Aiming to illustrate the history of English porcelain from its earliest manufacture to the mid-nineteenth century, this catalog describes 632 pieces in the Allen collection now owned by the Victoria and Albert Museum. The catalog also includes eighteen pieces of English earthenware and stoneware and several pages of photographs of the marks on the pieces in the collection.

_____. CATALOGUE OF ITALIAN MAJOLICA. 2 vols. Victoria and Albert Museum, Department of Ceramics. London: Board of Education, 1940. Vol. 1: xxiii, 485 p. Bibliography, pp. xix-xxiii. 2 indexes, pp. 458-85. Vol. 2: xxvii, 222 pls.

Grouping its subject according to locale and style, this catalog describes and illustrates some fifteen hundred examples of Italian majolica from the fourteenth to the eighteenth centuries in the collections of the Victoria and Albert Museum. Primitive, Florentine, Faenza, and Renaissance majolica; its Istoriato and Bianchi di Faenza types; tiles; and wares showing Chinese influence are all included.

Reynolds, Valrae. "Islamic Ceramics in the Newark Museum Collections." THE MUSEUM, n.s., no. 2 (1970): 1-27. 22 halftones. Bibliography, p. 27.

Using examples from the museum's collections, this issue of the Newark Museum's magazine briefly recounts the history of Islamic ceramics from the ninth to the eighteenth centuries. The illustrations of different types of ware and styles of decoration from Iran,

Turkey, Egypt, and Syria are accompanied by descriptions.

Salmony, Alfred. FAR EASTERN CERAMICS: THE CHARLES M. PRATT COLLECTION. Menasha, Wis.: George Banta Publishing Co. for the Vassar College Art Gallery, 1940. 79 p. 13 halftones.

Beginning with a brief history of the Pratt collection now in the possession of Vassar College, Salmony explains that the collection's emphasis is on Japanese pottery and porcelain although it includes a few examples of Chinese, Korean, and Southeast Asian ceramics. Very brief descriptions of the several hundred items in the collection make up the remainder of the catalog.

Sawyer, Alan R. ANCIENT PERUVIAN CERAMICS: THE NATHAN CUMMINGS COLLECTION. New York: Metropolitan Museum of Art, 1966; distributed by New York Graphic Society, Greenwich, Conn. 144 p. 2 color pls., 218 halftones and figs., maps. Bibliography, pp. 140-41. Index, pp. 142-44.

This is a well-illustrated and detailed popular history of the various cultures and changing styles of both north and south coast Peruvian ceramic art from its development around 1000 B.C. until the Inca Empire of 1500 A.D.

Solon, L.M. [Louis Marc]. CATALOGUE OF THE POTTERY AND PORCELAIN IN THE COLLECTION OF M.L. SOLON. Hanley, Stoke-on-Trent, Eng.: Charles Butters and Sons, 1912. 59 p. 25 pls.

This is the sales catalog of Louis Solon's collection of 683 pieces of pottery and porcelain. While the collection included some porcelains and European wares, most of it was composed of early English earthen and stonewares, which Solon had assembled to illustrate the history of the potteries at Staffordshire and to form a general ceramic history of England. Fifty to seventy-five pieces are illustrated in the plates.

Sullivan, Michael. CHINESE CERAMICS, BRONZES AND JADES IN THE COLLECTION OF SIR ALAN AND LADY BARLOW. London: Faber and Faber, 1963. 173 p. 164 pls. (4 color), 4 maps. Bibliography, pp. 165-70. Index, pp. 171-73.

This catalog illustrates and briefly identifies each piece in the Barlow collection, which contains works only from the Six Dynasties to the Ming periods. While it includes a number of bronzes and jades, the catalog is devoted principally to ceramics.

Tillyard, E[ustace].M[andeville].W[etenhall]. THE HOPE VASES: A CATALOGUE AND A DISCUSSION OF THE HOPE COLLECTION OF GREEK VASES WITH AN INTRODUCTION ON THE HISTORY OF THE COLLECTION AND ON LATE ATTIC AND SOUTH ITALIAN VASES. Cambridge: At the University Press, 1923. x, 179 p. 43 pls., figs. Index, pp. 173-78.

Since so much of this catalog's contents is adequately described in the book's title, it need only be added that all 357 pieces in the collection are identified and described and most are illustrated in the plates. The close cropping of many of the illustrations to show the paintings more clearly has unfortunately resulted in the loss of the shapes of the cropped vases.

Tracy, Berry Bryson. "English Pottery in the Collection of the Newark Museum." THE MUSEUM, n.s. 14, nos. 2-3, (1962): entire issue. 34 halftones. Bibliography, p. 30.

This issue of the Newark Museum's quarterly magazine presents a brief survey of the history of English pottery, excluding procelain, and is illustrated with examples from the museum's collection. The pieces date from the seventeenth to the mid-nineteenth centuries.

Trendall A[rthur].D[ale]. GREEK VASES IN THE LOGIE COLLECTION. Christchurch, N.Z.: University of Canterbury. 82 p. 41 pls. (1 color). Paper.

Beginning its introductory history of Greek vases with the Geometric style and ending with south Italian and Etruscan wares, Trendall's catalog identifies some sixty vases either in the Logie collection or lent to it. Each of the pieces is described in some detail and most are illustrated.

Walters, H[enry].B[eauchamp]. CATALOGUE OF THE ROMAN POTTERY IN THE DEPARTMENTS OF ANTIQUITIES, BRITISH MUSEUM. London: Printed by Order of the Trustees, 1908. liv, 464 p. 44 pls., figs. 4 indexes, pp. 449-64.

Introducing its subject with a history of these wares, this catalog identifies almost three thousand examples of pottery, particularly Roman, discovered in Britain. Although only a few dozen pieces are illustrated in the plates of this old volume, numerous stamps and ornamental patterns are reproduced.

Weaver, Muriel Porter. TLAPACOYA POTTERY IN THE MUSEUM COLLECTION. Indian Notes and Monographs, Miscellaneous Series no. 56. New York: Museum of the American Indian, Heye Foundation, 1967. xii, 48 p. 41 pls., figs., map. References, pp. 42-45. Paper.

This small study reports on the Tlapacoya pottery owned by the Museum of the American Indian. Weaver describes the site, cultural setting, and pottery types of this formative, or preclassical, period in the valley of Mexico. Vessels, figures, and a number of fragments are illustrated.

Chapter 10
CERAMIC MARKS

Numerous additional volumes contain chapters and/or appendixes on ceramic marks. They are noted in the bibliographical references throughout this volume.

*Barber, Edwin Atlee. MARKS OF AMERICAN POTTERS. 1904. Reprint. Southampton, N.Y.: Cracker Barrel Press, 1971. 174 p. 33 illus. Index, pp. 169-74. Paper. (Chamberlin, no. 1779).

Barber identifies one thousand marks from potteries in Ohio, Pennsylvania, New Jersey, New York, New England, and southern and other western states. His identifications are fairly detailed and occasionally include brief histories of the potteries and the marks of individual workers.

Bowes, James Lord. JAPANESE MARKS AND SEALS. London: Henry Sotheran & Co., 1882; Rpt. Ann Arbor, Mich.: Ars Ceramica, 1976. 379 p. Illus. Index, pp. 357-79. (Chamberlin, no. 1770).

The accuracy of parts of this volume, as of much early material on Japanese art, has been made questionable by contemporary scholarship. Nevertheless, Bowes's extensive text does contain information identifying hundreds of seals, marks, and characters. Part 1 of this three-part text deals with the marks on pottery and porcelain.

Callender, M.H. ROMAN AMPHORAE, WITH INDEX OF STAMPS. London: Oxford University Press, 1965. xxix, 323 p. 18 halftones in 7 pls., 27 figs., 4 maps. Bibliography, pp. xi-xviii. Index of stamps, pp. 57-280. Facsimiles of stamps, pp. 283-315.

Intended to fill the need for an index of stamps on amphorae found in western Europe, this text also includes brief essays relating the forms of the amphorae to their chronology and their functions.

Chaffers, William. THE COLLECTOR'S HAND-BOOK OF MARKS AND MONOGRAMS ON POTTERY AND PORCELAIN OF THE RENAISSANCE AND MOD-

ERN PERIODS. London: Reeves & Turner, 1893. xxiii, 178 p. Index, pp. vii-xxiii. (Chamberlin, no. 1743).

This condensed version of Chaffers's MARKS AND MONOGRAMS ON EUROPEAN AND ORIENTAL POTTERY AND PORCELAIN, (see next reference) is composed almost entirely of marks (3,000) and their identification. Details can be found in the larger work.

______. MARKS AND MONOGRAMS ON EUROPEAN AND ORIENTAL POTTERY AND PORCELAIN. 14th rev. ed. Edited by Frederick Litchfield. Los Angeles: Borden Publishing Co., 1946. xxviii, 1,095 p. Illus. Bibliography, pp. xvii-xxviii. Index, pp. 1061-95. (Chamberlin, no. 1744).

This is probably the most extensive and best-known book of ceramic marks in English. Although sections of this large work are devoted to the identification of Japanese and Chinese ceramics, the major portion is comprised of European marks.

Chait, Ralph M. A CHRONOLOGICAL ARRANGEMENT OF THE NAMES OF ALL THE CHINESE EMPERORS AND THE DATES OF THEIR REIGN TOGETHER WITH LISTS OF CHINESE SEALS, CHARACTERS AND SYMBOLIC MARKS. New York: Privately printed, 1928. 48 p. Bibliography, p. 47.

Because of its small size, this volume can deal with only a limited number of seals and characters.

Cushion, J[ohn].P., and Honey, W[illiam].B[owyer]. HANDBOOK OF POTTERY AND PORCELAIN MARKS. London: Faber and Faber, 1956. 476 p. Index, pp. 451-76. (Chamberlin, no. 1747).

Alphabetically arranged by country, which is subdivided into towns, examples of marks are included from virtually every European country. English marks are treated extensively; those of China and Japan only briefly. A number of nineteenth and twentieth century marks are also included.

Evans, Paul. ART POTTERY OF THE UNITED STATES: AN ENCYCLOPEDIA OF PRODUCERS AND THEIR MARKS.

See: Chapter 7: "Ceramics of the United States, Canada, and Mexico" (p. 107).

Hartman, Urban. PORCELAIN AND POTTERY MARKS. New York: 1942. 128 p. Index, pp. 125-28.

Beginning with brief histories of porcelain and several English companies, Hartman illustrates and identifies several hundred porcelain marks used in Europe, Russia, and America.

*Kovel, Ralph M., and Kovel, Terry H. DICTIONARY OF MARKS: POTTERY AND PORCELAIN. New York: Crown Publishers, 1953. 278 p. Bibliography, p. x. Index of marks, pp. 265-67. Index of manufacturers, pp. 268-78.

This book contains some 5,000 American, English, and European pottery marks and is conveniently organized by the prominent feature of each mark. Sections are arranged under each letter of the alphabet, except "Q," and under the various shapes and symbols used in the mark.

*Macdonald-Taylor, Margaret. A DICTIONARY OF MARKS: METALWORK, FURNITURE AND CERAMICS: THE IDENTIFICATION HANDBOOK FOR ANTIQUE COLLECTORS. New York: Hawthorn Books, 1962. 318 p. Bibliography, pp. 309-10. Index, pp. 311-18.

Although a few pages of the ceramics section of this dictionary are devoted to Japanese date marks and Chinese reign marks, the remainder of this section deals almost exclusively with English and European marks.

Ormsbee, Thomas H. ENGLISH CHINA AND ITS MARKS. Great Neck, N.Y.: Deerfield Editions, 1959. 200 p. Index, pp. 195-200.

While a few pages of this volume contain basic information about English china and pre-1750 English pottery, more than 150 of its pages compose the chapter entitled "Marks of English Porcelain, Earthenware and Stoneware, 1750-1880: Their Marks and Dates"; these marks are arranged alphabetically. Also included is a brief chapter devoted to present-day English china marks. Throughout the text the marks are the only illustrations.

Rhead, G. Woolliscroft. BRITISH POTTERY MARKS. London: Scott, Greenwood & Son, 1910. 303 p. 14 halftones.

Arranged clearly and alphabetically, Rhead's text briefly identifies hundreds of styles, companies, potteries, and individuals involved in the history of British ceramic production and illustrates 1,200 marks.

Thorn, C. Jordan. HANDBOOK OF OLD POTTERY AND PORCELAIN MARKS. New York: Tudor Publishing Co., 1947. 176 p. 44 illus. in 24 pls. Index, pp. 157-76.

Arranged alphabetically by country, this handbook contains several hundred marks from America and Europe. China, Japan, and the Near East are also included but only with a dozen or so marks from each. The plates here illustrate typical examples of the most famous wares.

Chapter 11

TECHNICAL WORKS ON CERAMIC MATERIALS AND PROCESSES

American Crafts Council. SALT GLAZE CERAMICS. New York: 1972. 36 p. Illus. Paper.

> This small exhibition catalog shows examples of historical and contemporary salt-glaze ware and discusses the process and pollution aspects of salt-glaze firing.

Ball, F. Carlton. SYLLABUS FOR ADVANCED CERAMICS. Bassett, Calif.: Keramos Books, 1972. 68 p. Charts. Index, pp. 66-68. Paper.

> Similar in format and scope to Ball's SYLLABUS FOR BEGINNING POTTERY (see next reference), this syllabus concentrates on some of the more technical aspects of ceramic glazes. Information on clay bodies and kiln firings is also included.

______. SYLLABUS FOR BEGINNING POTTERY. Bassett, Calif.: Keramos Books, 1971. 76 p. Figs., charts. Glossary, pp. 58-69. Index, pp. 73-76. Paper.

> Designed for use in schools, this book provides a clearly organized introduction to ceramics for the classroom or a supplement to the more common texts on beginning pottery. It does not duplicate these very basic texts but offers a sound introduction to the more technical aspects of making pottery.

Ball, F. Carlton, and Lovoos, Janice. MAKING POTTERY WITHOUT A WHEEL. New York: Van Nostrand Reinhold Co., 1965. 160 p. 197 halftones. Glossary, pp. 154-57. Bibliography, p. 158. Index, p. 159.

> Primarily concerned with promoting a sense of texture and form in the making of pottery, Ball introduces the reader to the methods and ideas that have inspired his own ceramic work. He demonstrates, step-by-step, the hand building of an extensive range of forms, and there are excellent illustrations of his textures, techniques, and finished pieces. A brief chapter on glazes and kilns is also included.

Barber, Edwin Atlee. THE CERAMIC COLLECTOR'S GLOSSARY.

See: Chapter 2: "General Histories, Dictionaries, and Encyclopedias" (p. 7).

Behrens, Richard. GLAZE PROJECTS: A FORMULARY OF LEADLESS GLAZES. A CERAMICS MONTHLY Magazine Handbook, edited by Thomas Sellers. Columbus, Ohio: Professional Publications, 1971. 64 p. Tables, formulas. Glossary, pp. 59-62. Paper.

Behrens enumerates a large number of glaze formulas for different clay bodies, colors, and temperatures. Formulas for special glazes, such as Egyptian paste, Parian ware, raku, terra sigillata, and cement are included.

Bellaire, Marc. UNDERGLAZE DECORATION. A CERAMICS MONTHLY Magazine Handbook, edited by Louis G. Farber. Columbus, Ohio: Professional Publications, 1957. 64 p. Halftones. Paper.

Intended for the beginning hobbyist, this introductory text explains a variety of underglaze techniques and includes a number of sample designs for the beginner to try.

Berensohn, Paulus. FINDING ONE'S WAY WITH CLAY. New York: Simon & Schuster, 1972. 159 p. Illus. 8 color pls., halftones. Glossary, pp. 157-58. Bibliography, p. 159.

This excellent study offers an extensive explanation of the pinching method with clay and an intensely personal presentation of one person's creative growth. Discussing various approaches to the pinching method, the coloring and inlaying of clays, and the use of sawdust kilns, the book is handsomely illustrated throughout with examples of the author's work.

*Billington, Dora M. POTTERY FOR EVERYONE. Revised by John Colbeck. New York: Watson-Guptill Publications, 1974. 175 p. 240 halftones, tables. Glossary, pp. 164-68. Bibliography, pp. 168-69. Index, pp. 173-75.

This revised edition of Billington's THE TECHNIQUE OF POTTERY (see next entry) includes descriptions of basic pottery-making methods, as well as technical information on glazing, firing, and various decorative techniques. These include the use of agate ware, slip trailing, marbling, sgraffito, lusters, and underglaze and overglaze painting.

______. THE TECHNIQUE OF POTTERY. London: B.T. Batsford, 1962. 222 p. Illus., charts. Glossary, pp. 200-212. Index, pp. 217-22.

Designed as a text for the serious student, this volume covers the basic techniques of working with clay, including hand building, throwing, glazing, and firing. This work was revised in 1974 as

POTTERY FOR EVERYONE (see preceding entry).

Binns, Charles F. THE POTTER'S CRAFT. 4th ed., revised by John F. McMahon. Princeton, N.J.: D. Van Nostrand Co., 1967. 144 p. Illus. 4 appendixes, pp. 124-33. Glossary pp. 134-40. Index, pp. 141-44.

Although quite brief and containing a limited number of illustrations, this book does cover a good deal of basic technical information for the serious student of ceramics. The text includes a very short history of Eastern and Western ceramics and information on clays, bodies, glazes, forming methods, and mold making.

Cardew, Michael. PIONEER POTTERY. London: Longman Group, 1969. xv, 327 p. 57 halftones in 24 pls., figs., charts. 15 appendixes, pp. 255-309. Bibliography, pp. 310-11. Index-Glossary, pp. 313-27.

Intended as a practical aid to potters working in remote areas, Cardew's book is an extensive and valuable study for both the beginner and the experienced craftsman. It is filled with technical information and the personal experience of the author's many active years. Its emphasis is on independence and personal discovery and a good deal of space is given to understanding raw materials, tools, and practical techniques. A number of technical volumes are cited in the bibliography. This reference text would be a valuable addition to a craftsman's library.

CERAMICS FOR SECONDARY SCHOOLS. Curriculum Bulletin, 1961-62 series, no. 7. Brooklyn: Board of Education of the City of New York. 130 p. Illus. Paper.

This curriculum bulletin is designed to supplement the syllabus in the New York City Board of Education's publication INDUSTRIAL ARTS FOR GRADES SEVEN, EIGHT AND NINE. Although somewhat dated, the bulletin does contain a good deal of basic information that could prove useful to the nonspecialist who may be teaching a course at some time in the ceramic field.

Clark, Kenneth. PRACTICAL POTTERY AND CERAMICS. New York: Viking Press, 1964. 80 p. Illus. Bibliography, p. 6. Glossary, p. 77. Conversion tables, p. 78. Index, pp. 79-80.

This book provides the beginner with an introduction to the ceramic process, but some of the basic areas touched upon are treated with no detail whatsoever. The strongest aspect of the book lies in its coverage of contemporary ceramic design and the majority of illustrations are most valuable because they show that design.

*Colbeck, John. POTTERY: THE TECHNIQUE OF THROWING. New York: Watson-Guptill Publications, 1969. 159 p. Index, p. 159.

If throwing can be learned from a book, Colbeck's excellent volume is probably the most complete source available. Devoted entirely to the subjects of throwing and turning (trimming), the book is fully illustrated and carefully describes each step of the various throwing situations--centering, opening, forming, making lids, use of tools, large-scale work, combining pieces, and joining.

Conrad, John W. CERAMIC FORMULAS: THE COMPLETE COMPENDIUM: A GUIDE TO CLAY, GLAZE, ENAMEL, GLASS AND THEIR COLORS. New York: Macmillan Publishing Co.; London: Collier Macmillan Publishers, 1973. 309 p. Color chart, 9 figs. Bibliography, p. 309.

In addition to providing some 700 tested formulas, this handbook introduces and advises methods of testing each of the subjects named in its title. Clay and glaze formulas ranging from the lowest earthenware (cone 022) to the highest fired porcelain (cone 13+) are all represented.

Counts, Charles. COMMON CLAY. Limited ed. Atlanta: Droke House/Hallux, 1971. 144 p. 16 color pls. Halftones.

This is a very personal book in which the author combines poetry and philosophy to praise all potters and their craft. Counts talks about his own work but focuses his text on the traditional work of eight folk potters in the American South.

_____. POTTERY WORKSHOP: A STUDY IN THE MAKING OF POTTERY FROM IDEA TO FINISHED FORM. New York: Macmillan Co.; London: Collier Macmillan Publishers, 1973. 198 p. 293 illus. Bibliography, pp. 195-96. Index, pp. 197-98.

With an informal and personal point of view, this well-illustrated book covers the basic range of clay work, from digging one's own clay to the marketing of the finished product. Diagrams for building a wheel and kiln and information on glaze chemicals and firing procedures are also included.

Cox, George J. POTTERY: FOR ARTISTS, CRAFTSMEN AND TEACHERS. Technical Art Series, edited by Paul Monroe. New York: Macmillan Co., 1914. 200 p. 74 figs. Glossary, pp. 185-92.

This very old technical manual includes general information on forming methods, making glazes, and firing.

Coyne, John, ed. THE PENLAND SCHOOL OF CRAFTS BOOK OF POTTERY. A Rutledge Book. Indianapolis and New York: Bobbs-Merrill, 1975. 192 p. Illus. (part color). Glossary, p. 192.

Handsomely illustrated with photographs of the artists at work and of their completed projects, this volume describes in the words of each of eight potters--Tyrone and Julie Larson, Jane Peiser, Tom Suomalainen, Bruno LaVerdiere, Cynthia Bringle, Toshiko Takaezu, Don Reitz, and Ron Propst--a project illustrative of his or her work. Each describes his or her philosophy and feelings as well as his or her working techniques. Brief essays on hand building by Paulus Berensohn and on throwing by Robert Turner, as well as lists of formulas and sources of supplies, are also included.

Crespi, David E. CERAMIC GLAZE FORMULAS, Book 1. New Haven, Conn.: Southern Connecticut State College, n.d. 52 p. Index. Paper.

This pamphlet is the result of extensive glaze experimentation carried on by undergraduate art majors in the ceramics department at Southern Connecticut State College. It includes an identification of ceramic materials and a variety of glaze formulas.

______. "Experimental Pottery Techniques for the Junior and Senior High School Art Program." New Haven: Southern Connecticut State College, 1972. 30 p. Illus. Mimeographed.

Crespi describes a series of simple hand-building techniques requiring a minimum of space and equipment. The techniques include paddle wedging, drape construction, paddle construction, plaster molds, sand molds, and sling construction.

Curtis, Edmund de Forest. POTTERY: ITS CRAFTSMANSHIP AND ITS APPRECIATION. New York and London: Harper & Brothers, Publishers, 1940. 100 p. 19 pls. Bibliography, pp. 94-97. Index, pp. 99-100.

Intended as a technical handbook for those seriously interested in good pottery rather than simply in handicraft, this volume contains some basic information on clay bodies, glazes, and forming and firing techniques. Although only an introduction and lacking any real detail, Curtis's text does manage to convey throughout, the author's strong sense of appreciation.

Dalton, W.B. CRAFTSMANSHIP AND DESIGN IN POTTERY. London: Sir Isaac Pittman & Sons, 1957. 113 p. 85 illus. Bibliography, p. 91. Glossary, pp. 93-98. Appendixes, pp. 99-110. Index, pp. 111-13.

Dalton's book is about equally divided between a historical survey of ceramics and a discourse on studio work. The book's latter part places special emphasis on glaze experimentation.

Dickerson, John. RAKU HANDBOOK: A PRACTICAL APPROACH TO THE CERAMIC ART. New York: Van Nostrand Reinhold Co.; London: Studio Vista, 1972. 112 p. 91 illus. (40 figs.). Suppliers, pp. 106-7. Bibliography, p. 108. Glossary, pp. 109-12.

Its brief history and illustrated explanation of traditional Japanese techniques and its extensive technical information on clay bodies, glazes, and kilns make this a valuable handbook for anyone seriously interested in making raku ware.

Doat, Taxile. GRAND FEU CERAMICS: A PRACTICAL TREATISE ON THE MAKING OF FINE PORCELAIN AND GRES.

See: Chapter 6: "Western Ceramics: The Middle Ages to the Twentieth Century" (p. 89).

Dodd, A[rthur].E[dward]. DICTIONARY OF CERAMICS. New York: Philosophical Library, 1964. 327 p. 3 appendixes, pp. 320-27.

Written by the information officer of the British Ceramic Research Association, this dictionary covers the whole range of ceramic work. Strongly emphasizing the more technical terminology of industrial and scientific ceramics, the dictionary also provides widely used trade names and American usage.

Drake, Kenneth. SIMPLE POTTERY. New York: Watson-Guptill Publications; London: Studio Vista, 1966. 96 p. Illus. (incl. 4 color pls., figs.). Bibliography, p. 90. Index, pp. 95-96.

This small book includes the usual introductory information on clay, equipment, and hand-building techniques. However, it also includes an explanation of plaster- and mold-making techniques and some information on glazing and firing.

Duncan, Julia Hamlin, and D'Amico, Victor. HOW TO MAKE POTTERY AND CERAMIC SCULPTURE: TWENTY GRADED PROJECTS. Art for Beginners series. New York: Museum of Modern Art, 1947; distributed by Simon and Schuster, New York. 95 p. Halftones. Bibliography, p. 92. Index, pp. 93-94.

Published as an outgrowth of the Museum of Modern Art's program for the rehabilitation of war veterans, this book clearly illustrates the step-by-step completion of several simple projects.

*Farnsworth, Warren. BEGINNING POTTERY. New York: Van Nostrand Reinhold Co., 1973. 128 p. 215 illus. Bibliography, p. 127.

Designed as a teaching aid in introducing clay to younger children, this short, attractive volume includes information on the creative use of clay despite minimum facilities and equipment. Information on either buying or digging one's own clay and firing with or without a kiln is also included.

Ford, Betty Davenport. CERAMIC SCULPTURE. New York: Reinhold Publishing Corp., 1964. 100 p. Illus. (part color), figs. Bibliography, pp. 98-99. Glossary, p. 99. Index, p. 100.

This introduction to the techniques of ceramic sculpture discusses the design, construction, firing, and finishing of work. Some technical information on clay bodies, firing procedures, and glazes; suggestions for class projects; and a gallery of works done by the author are included.

Fournier, Robert. ILLUSTRATED DICTIONARY OF PRACTICAL POTTERY. New York: Van Nostrand Reinhold Co., 1973. 256 p. Illus. (incl. figs.), diagrams. Bibliography, p. 254.

This illustrated dictionary clearly defines numerous technical terms for describing working with clay. The book explains modeling techniques, the formation of glazes, and kiln firing and identifies materials, tools, and equipment.

Fraser, Harry. ELECTRIC KILNS. Edited by Sarah Bodine. New York: Watson-Guptill Publications, 1974. 144 p. 71 halftones, figs., diagrams, tables. Bibliography, p. 140. Suppliers list, p. 141. Glossary and index, pp. 142-44.

Although it focuses on electric kilns, this text briefly describes the evolution of kilns in general and discusses pottery faults and experimental glazing. Its sections on electric kilns include information on purchasing and building a kiln, the use of accessories, firing procedures, and the effects of heat on clay and glazes. Fraser also reprints the entire text of Lawrence Webber's MAKING STUDIO ELECTRIC KILNS (see p. 172).

_____. GLAZES FOR THE CRAFT POTTER. London: Pitman Publishing, 1973; New York: Watson-Guptill Publications, 1974. 146 p. 62 halftones and figs., tables. 11 appendixes, pp. 127-41. Bibliography, p. 142. Index, pp. 143-46.

While this book does include a few pages of sample glaze formulas, it focuses on a more technical understanding of glaze materials and their use. Fraser examines the chemistry of glazes, including the effects of heat; health precautions; pottery faults; and a variety of additional technical information on glaze formation and application.

Furnival, William James. LEADLESS DECORATIVE TILES, FAIENCE AND MOSAIC . . . IN THE HISTORY, MATERIALS, MANUFACTURE AND USE OF ORNAMENTAL FLOORING TILES, CERAMIC MOSAIC AND DECORATIVE TILES AND FAIENCE.

See: Chapter 2: "General Histories, Dictionaries, and Encyclopedias" (p. 10).

Goldberg, Steven A. GLAZE CALCULATION. San Jose, Calif.: Billiken Press, 1970. 68 p. Tables, charts. Paper.

Intended as a source book for the independent potter and as a textbook for classroom use, Goldberg's publication introduces the student to the technical problems of glaze chemistry using a step-by-step approach. Central to Goldberg's treatment is the accurate conversion of glaze recipes from an empirical formula to a "batch" formula. While its subject is not light, the material is presented in a readable and straightforward manner.

Grebanier, Joseph. CHINESE STONEWARE GLAZES. New York: Watson-Guptill Publications; London: Pitman Publishing, 1975. 144 p. 16 color pls., figs., tables. Appendix, pp. 132-34. Glossary, pp. 135-40. Bibliography, p. 141. Index, pp. 142-44.

Grebanier describes the processes of reproducing several of the famous Chinese stoneware glazes, which date mostly from the Sung dynasty. He treats in detail the techniques and formulas for producing light blue Chun, the celadons, blue-and-brown mottled glazes, copper red, peach bloom, *temmoku*, and *Tz'u-chou*. Since most of these are high fire (cone 10), reduction (gas, oil) glazes, this material is not really useful to the novice or the hobbyist with limited facilities. Information on new glazes developed by the author, clay bodies, slips, glaze calculation, and synthetic wood ash are also supplied.

Hamer, Frank. THE POTTER'S DICTIONARY OF MATERIALS AND TECHNIQUES. London: Pitman Publishing; New York: Watson-Guptill Publications, 1975. x, 349 p. 554 halftones, figs. Appendix of tables, p. 327-48. Bibliography, p. 349.

Intended for use by potters, teachers, and students, this comprehensive volume is alphabetically organized as an index of key words and phrases to explain the nature and behavior of ceramic materials, techniques, processes, and historical types. Complex subjects are treated in some detail and many of the entries are cross-referenced for further information. Some thirty tables of chemical and technical information are contained in the appendix.

*Hamilton, David. MANUAL OF POTTERY AND CERAMICS. New York: Van Nostrand Reinhold Co., 1970. 188 p. 170 illus. (part color and figs.). 3 appendixes, pp. 175-80. Glossary, pp. 181-83. Bibliography, p. 183. Index, pp. 184-88.

Aimed at the first-year student or beginner, Hamilton's text contains a very short history of ceramic art and describes each of its basic processes, from hand forming to firing. Clay preparation, types of kilns, and glazes are treated clearly, but without a great deal of technical detail. The illustrations are of antique and contemporary pieces, including works by the author.

Hartung, Rolf. EXPLORING CLAY: HAND TECHNIQUES. New York: Van Nostrand Reinhold Co., 1972. 79 p. Illus.

With an absolute minimum of text, Hartung illustrates the forming of a number of simple, practical, yet pleasing objects with one's hands and a few other basic tools.

Harvey, Roger; Kolb, Sylvia; and Kolb, John. BUILDING POTTERY EQUIPMENT. New York: Watson-Guptill Publications; London: Pitman Publishing, 1975. 208 p. Illus., tables. Index, p. 206.

Aiming to make available to the reader equipment that is of high quality, yet economical and suited to the needs of the individual craftsperson, the authors have included plans and descriptions of clay preparation equipment, wheels, hand tools, kilns, and firing and other workshop equipment. The text is full of clear illustrations and a good deal of technical information on the construction, sources, and materials of ceramic equipment.

Hetherington, A[rthur].L[onsdale]. CHINESE CERAMIC GLAZES.

See: Chapter 4: "Eastern Ceramics" (p. 29).

Hofsted, Jolyon. STEP-BY-STEP CERAMICS. New York: Golden Press, 1967. 95 p. Illus. Glossary, p. 93. Bibliography, p. 95. Suppliers, p. 95.

Designed for the beginning potter and heavily illustrated in color and black and white, this volume briefly covers each of the basic ceramic techniques. Space is also devoted to studio organization, glazing, building and firing a kiln, raku, and a portfolio of projects.

Home, Ruth M. CERAMICS FOR THE POTTER. Peoria, Ill.: Charles A. Bennett Co., 1953. 229 p. Illus. Bibliography, pp. 207-25. Index, pp. 227-29.

Home's book contains a good deal of technical information for the beginning but serious potter. The book is not very well illustrated but does contain useful charts and a long bibliography. The emphasis of the text is on clay bodies and the history of various potting techniques and tools. Glass and enamels are also covered.

Honore, York. POTTERY MAKING: FROM THE GROUND UP. New York: Viking Press, 1941. 72 p. Illus.

Although it is not particularly detailed, this simple text explains, with as little dependence as possible on outside sources, the making of ceramic pieces. A few fundamental forming techniques and an introduction to wheel throwing are included, but the chapters on digging one's own clay, making bricks, simple glazing, and building one's own kiln and potter's wheel are this book's real strengths.

Iwao, Shuichi, editor in chief. THE CLAYS OF JAPAN. [Kawasaki], Jap.: Geological Survey, 1969; Tokyo: International Clay Conference, 1969. 209 p. Illus., tables, charts.

This collection of lectures sponsored by the Science Council of Japan and delivered in Tokyo describes representative clay deposits and the utilization of various economically important clays. Recent investigations in clay minerology by Japanese scientists are also described. This material is scientifically oriented and quite technical.

Johnson, William H., and Newkirk, Louis V. THE CERAMIC ARTS. New York: Macmillan Co., 1942. 158 p. Illus. Index, pp. 157–58.

Intended as an industrial arts text for grades seven, eight, and nine, this very dated book includes chapters on pottery, plastics, glass, alabaster carving, and cement. None of these subjects are dealt with in real detail.

*Jorgensen, Gunhild. THE TECHNIQUES OF CHINA PAINTING. Translated by Hugh Young. New York: Van Nostrand Reinhold Co., 1974. 112 p. Illus. (part color and figs.). Index, p. 112.

This hobbyist's primer begins with an explanation of the materials, equipment, and painting fundamentals used to create different types of wares. The book's second part deals with slightly more advanced techniques and illustrates a variety of decorations that can be used.

Kenny, John B. CERAMIC DESIGN. Philadelphia: Chilton Books Co., 1963. 322 p. Illus. (incl. 76 color, figs.). Glossary, pp. 316-19. Index, pp. 320-22.

Kenny's fully illustrated text examines the designing, forming, and various methods of decorating clay objects. Carving, drawing, glazing, firing, and the use of color are some of the decorating methods the book treats.

*______. CERAMIC SCULPTURE. New York: Greenberg, Publisher; Toronto: Ambassador Books, 1953. 304 p. 80 illus. Index, pp. 303–4.

Although dated in its appearance, Kenny's book covers a broad range of the techniques and skills needed for the production of ceramic sculpture. Several chapters are devoted to tools, materials, colors, and kilns.

______. THE COMPLETE BOOK OF POTTERY MAKING. Philadelphia: Chilton Book Co., 1949. 242 p. Pls. (incl. 4 color), figs. Index, pp. 239–42.

Although looking somewhat dated, this volume covers the basic range of clay work and includes fairly detailed information on hand building, using the potter's wheel, glazing, and firing. A good deal of space here is also devoted to molds and slipcasting.

Keyser, Mrs. C. Naaman. A HOME CRAFT COURSE IN POTTERY. 2d ed. The Home Craft Course Series, vol. 1. Plymouth Meeting, Pa.: 1945. 40 p. Figs. Paper.

One of about twenty volumes by Keyser, this small introduction to clay work gives rudimentary information on clay techniques and decoration. While the book probably fulfilled a clear need when it was published, its contents are covered in more detail and with greater clarity by any one of several more recent "how-to" books.

______. PENNSYLVANIA GERMAN POTTERY. The Home Craft Course Series. Plymouth Meeting, Pa.: 1943. 36 p. Pls. (1 color), figs. Paper.

Keyser's text very briefly describes the techniques of making traditional Pennsylvania German pottery. Most of this pamphlet, however, is devoted to the figure illustrations of traditional designs used on these wares. While they give the reader the general idea of what the designs look like, these illustrations are not particularly clear nor detailed.

Kingery, W.D., ed. CERAMIC FABRICATION PROCESSES. Cambridge, Mass.: Technology Press of Massachusetts Institute of Technology; New York: John Wiley & Sons, 1958. 235 p. Illus., tables, charts. Indexes, pp. 229-35.

This volume is a collection of lectures delivered at a special summer program in ceramics at M.I.T. during 1956 and not intended as a definitive text. Attempting to evaluate the importance of current developments in ceramic processing, this material tends to be quite technical and basically deals with mass production and industrial ceramics.

Kriwanek, Franz. CERAMICS IN TEXAS: USES OF TEXAS CLAYS AND LOCAL MATERIALS IN SMALL SCALE PRODUCTION AND EDUCATIONAL CERAMICS. Lubbock, Tex.: Texas Tech Press, 1968. 48 p. Illus. Paper.

Intended to supplement regular ceramic texts, this pamphlet furnishes information on local resources in western Texas. Much of this material on the utilization of local resources, however, seems readily adaptable to almost any area.

Krum, Josephine R. HAND-BUILT POTTERY. International Textbooks in Art Education, Italo L. deFrancesco, consulting ed. Scranton, Pa.: International Textbook Co., 1960. 116 p. Illus. Bibliography, pp. 112-13. Index, pp. 114-16.

Rather than a detailed technical manual, this volume is a more personal attempt by the author to share her experience of clay as a creative medium in the classroom. Technical information is included, but the emphasis is on exploration by both student and teacher.

Lawrence, W[illis].G[rant]. CERAMIC SCIENCE FOR THE POTTER. Philadelphia: Chilton Book Co.; Don Mills, Ontario: Thomas Nelson & Sons, 1972. xiii, 240 p. Charts, diagrams, figs. 7 appendixes, pp. 219-34. Index, pp. 235-39.

Although not intended as a profound scientific effort, this is a fairly technical examination of the nature of clays and glaze materials written for the artist-potter. The book deals with kiln pollution, terra sigillata, and lead glazes and contains further reference material.

Leach, Bernard. A POTTER'S BOOK. 2d ed. London: Faber and Faber, 1940. 294 p. Pls. (incl. 4 color), figs., charts. Appendix, pp. 261-83. Index, pp. 284-94. (Chamberlin, no. 1783).

Dealing with English slipware, stoneware, oriental porcelain, and Japanese raku, Leach's text contains a good deal of valuable technical information and conveys a strong personal feeling for good craftsmanship. This is an excellent reference book for any serious student of ceramics.

LEAD GLAZES FOR DINNERWARE. ILZRO Manual, Ceramics, no. 1. New York: International Lead Zinc Research Organization, 1971. 142 p. Charts. Paper.

A combined effort of the International Lead Zinc Research Organization and the United States Potters Association of East Liverpool, Ohio, this manual is the product of a research program to develop sound data on the proper formulation and firing temperatures required to minimize the release of lead from lead-containing glazes.

LeFevre, Leon. ARCHITECTURAL POTTERY. Translated by K.H. Bird and W. Moore Binns. London: Scott, Greenwood & Co., 1900. xv, 496 p. 5 pls., 950 figs., tables. Bibliography, pp. 485-86. Index, pp. 492-96.

This introduction to the technical processes of industrial pottery briefly explains each facet of working with clay, from the properties of the materials themselves to their preparation and numerous uses. The manufacture and use of bricks, tiles, pipes, terracottas, glazes, decorated quarries, and architectural and sanitary pottery are all described.

Long, Lois Culver. CERAMIC DECORATION. Indianapolis: American Art Clay Co., 1958. 59 p. Halftones. Glossary, p. 55. Bibliography, p. 56. Index, pp. 58-59. Paper.

This introduction briefly covers a variety of methods for decorating ceramic ware. Included are ideas for working with the clay itself, the use of slips, underglazes and overglazes, and a variety of glazing techniques.

Luisi, Billie. POTWORKS: A FIRST BOOK OF CLAY. New York: William Morrow & Co., 1973. xxix, 154 p. Figs. Word list, pp. xiii-xxvii. Annotated book list, pp. 141-45. Suppliers, pp. 147-49. Index, pp. 151-54.

This imaginative and readable text is a simple but informed introduction to the making of one's own pots. Details have been kept to a minimum and emphasis is on the establishment of one's own studio through hard work and creativity rather than a large financial outlay.

Lynngaard, Finn. POTTERY: RAKU TECHNIQUE: DESIGN, MATERIAL, TECHNIQUE. New York: Van Nostrand Reinhold Co., 1970. 79 p. Illus. Glossary, pp. 73-77. Bibliography, p. 78. Index, p. 79. Paper.

Although this is a very small volume, it manages to incorporate more information on the history and making of raku pottery than is sometimes covered in much larger texts. In addition to a summary of the Japanese raku tradition and contemporary American work, the book provides technical information on clay bodies, glazes, and kiln construction.

Mann, Kathleen. CHINA DECORATION. New York: Pitman Publishing Corp., 1953. 47 p. 7 pls. (1 color), figs. Paper.

This short introduction covers the basic technical information on china decoration and includes line drawings and reproductions of finished pieces. This author's approach encourages imagination rather than copying and stresses the importance of good design.

March, Benjamin. STANDARDS OF POTTERY DESCRIPTION.

See: Chapter 2: "General Histories, Dictionaries, and Encyclopedias" (p. 12).

Mavros, Donald O. GETTING STARTED IN CERAMICS. Getting Started Series. New York: Bruce Publishing Co.; London: Collier Macmillan, 1971. 84 p. 77 illus. (10 color). Glossary, pp. 77-82. Index, pp. 83-84. Paper.

This is a very brief but extensively illustrated volume, which includes some material on the history of ceramics, forming methods, glazing, and firing.

Memmott, Harry. THE ART OF MAKING POTTERY: FROM CLAY TO KILN. New York: Clarkson N. Potter, 1971; distributed by Crown Publishers. 251 p. 450 illus. Glossary, pp. 234-44.

This profusely illustrated, introductory text covers several basic areas of ceramic work for the beginning potter: the techniques of throwing, hand building, molding, various methods of decoration, glazing, and the firing of kilns.

*Meyer, Fred. SCULPTURE IN CERAMICS. New York: Watson-Guptill Publications, 1971. 157 p. Illus. Bibliography, p. 151. List of suppliers, pp. 153-54. Index, pp. 155-57.

Covering a wide range of information of value to a student of ceramic sculpture, this book includes a brief and highly personal historical survey, an explanation of clay types, kilns, tools, and glazes and even discusses the formal presentation of a completed work. Three heavily illustrated chapters are devoted to the various sculptural techniques used by the author, his wife (Joan Frantz Meyer), and Hobart Cowles.

Miserez-Schira, Georges. THE ART OF PAINTING ON PORCELAIN. Translated by Camilla Sykes. Radnor, Pa.: Chilton Book Co., 1974. 133 p. Illus. (part color and figs.). Glossary, pp. 126-29.

The author provides a step-by-step guide for the beginning painter, from arranging the work space to preparing the paints and colors, from describing the painting of the simplest flower to mastering birds and complex landscapes. The author also offers a brief review of the history of ceramics, instructions on tile painting, and ideas for the more experienced painter. Handsomely illustrated, this book often shows examples of very elegant and sophisticated designs.

Nagumo, Ryu. JAPANESE POTTERY AS A HOBBY. Tokyo: Toto Shuppan Co.; Tokyo and Rutland, Vt.: Japan Publications Trading Co., 1963. 128 p. 427 illus. (10 color, 27 figs.), charts.

While the main part of this well-illustrated text covers a variety of traditional Japanese hand-building techniques, space is devoted to a good deal of useful technical information on the building and firing of kilns, glazing, and various formula tables.

*Nelson, Glenn C. CERAMICS: A POTTER'S HANDBOOK. 3d ed. New York: Holt, Rinehart & Winston, 1971. 348 p. Illus., tables, charts. Appendix, pp. 303-24. Bibliography, pp. 329-32. Glossary, pp. 333-39. Index, pp. 341-48.

This comprehensive studio handbook contains much valuable information for the serious student and working potter. Beginning with a brief history and a chapter on contemporary world ceramics, the text primarily covers technical information on clays, glazes, kilns, and equipment. Several pages of glaze and clay formulas are located in the appendix.

Newcomb, Rexford. CERAMIC WHITEWARES: HISTORY, TECHNOLOGY AND APPLICATIONS. New York: Pitman Publishing Corp., 1947. 313 p. Illus., tables. Appendix, pp. 303-8. Index, pp. 309-13.

> Identifying "whitewares" as various types of pottery, porcelain, and china "no matter what the end use," the author is principally concerned with the engineering properties and possible uses of these wares. Forming processes, firing, glazes, electrical applications, industrial uses, and abrasives are all discussed.

Norton, F[rederick].H[arwood]. CERAMICS: AN ILLUSTRATED PRIMER. Garden City, N.Y.: Hanover House, 1960. 92 p. Illus. Glossary, pp. 84-89. Index, pp. 90-92.

> This revised edition of the first part of Norton's CERAMICS FOR THE ARTIST POTTER (see next entry) briefly covers the most basic areas of ceramic work for the hobbyist and beginning student.

_____. CERAMICS FOR THE ARTIST POTTER. Cambridge, Mass.: Addison-Wesley Publishing Co., 1956. 320 p. Illus., charts, diagrams. Glossary, pp. 302-8. Appendix, pp. 309-12. Index, pp. 313-20.

> While somewhat dated in appearance, this book does embrace a good deal of technical information on clay bodies, glazes, decoration, and firings. It also covers historical development and principles of design.

_____. ELEMENTS OF CERAMICS. Addison-Wesley Metallurgy Series; Morris Cohen, consulting ed. Reading, Mass.: Addison-Wesley Publishing Co., 1952. 247 p. Figs., charts, tables. Index, pp. 241-47.

> Written by a professor of ceramics at the Massachusetts Institute of Technology, this volume is intended as an introduction to ceramic science for undergraduate students and ceramic artists seeking a more thorough knowledge of their field. Well illustrated with charts and diagrams, Norton's text covers a good deal of technical information on the nature of clays, glaze materials, and kiln firing.

Olsen, Frederick L. THE KILN BOOK. Bassett, Calif.: Keramos Books, 1973. xii, 146 p. Illus., charts, tables. Appendix, pp. 121-44. Index, pp. 145-46. Paper.

> This volume contains a great deal of technical information on kilns, their principles of design, and materials and methods of construction. The book also contains a good selection of illustrations and information on the various types of kilns, their different styles and purposes, and fuels (including electricity). Although the author includes all the technical information necessary for building one's own kiln, the text is enjoyably informal with the author explaining the various designs of kilns that he has either seen or helped build.

Parmelee, Cullen W. CERAMIC GLAZES. 2d ed. Edited by E.D. Lynch and H.L. Friedberg. Chicago: Industrial Publications, 1951; distributed by Cahners Book Division, Boston. 322 p. Illus., charts, diagrams, tables. Index, pp. 315-22.

This thorough, technical volume on ceramic glazes includes information on their materials, formulation, special types, and defects. This fairly detailed text is for the serious student and the professional craftsperson.

Petterson, Henry. CREATING FORM IN CLAY. New York: Reinhold Book Corp., 1968. 112 p. 200 illus. Glossary, pp. 104-11.

Designed primarily for older schoolchildren and their teachers, Petterson's book demonstrates a variety of useful hand-building techniques and conveys a strong sense of exploration in design. Methods of decorating, glazing, and firing are also covered.

Piepenburg, Robert. RAKU POTTERY. New York: Macmillan Co.; London: Collier Macmillan, 1972. 159 p. Illus. (part color and figs.). Bibliography, p. 155. Index, pp. 157-59.

Well illustrated and containing a good deal of technical information on raku clay, glazes, and firing, this volume also includes a history of Japanese raku and the tea ceremony. Throughout his text Piepenburg warmly expresses the spirit and philosophy of his work.

Priolo, Joan B. CERAMICS AND HOW TO DECORATE THEM. New York: Sterling Publishing Co., 1958. 144 p. Illus. (incl. 8 color).

This is a beginner's book for the ceramic hobbyist. The text includes chapters on molds, under- and overglaze decoration, and copper enameling. While some of its technical information may be useful, this text is badly dated and suffers from a very weak sense of design.

*Pucci, Cora. POTTERY: A BASIC MANUAL. Little, Brown Craft Series. Boston: Little, Brown & Co., 1974. 171 p. Illus. Bibliography, p. 171.

Writing informally for the beginner, Pucci describes the basic methods of working with clay--hand building and the potter's wheel--and introduces the novice to glazing, decoration, firing, and studio organization. She includes some technical information, but her major goals are to provide encouragement, to inspire experimentation, and to enable her readers to experience the joy of personal expression.

Rado, Paul. AN INTRODUCTION TO THE TECHNOLOGY OF POTTERY. The Commonwealth and International Library of Science, Technology, Engineering and Liberal Studies. Oxford: Pergamon Press, 1969. xx, 258 p. Illus. (part color and figs.). 2 appendixes, pp. 217-29. References, pp. 231-39. Glossary, pp. 241-44. Index, pp. 245-58. Paper.

> This handbook briefly introduces the reader to a broad range of material on ceramic technology. Rado outlines the development of pottery, the history of the industry, and the variety of pottery types. Much of the text is devoted to brief descriptions of the chemistry of clay and glaze materials, calculations, and mechanical processes.

Rhodes, Daniel. CLAY AND GLAZES FOR THE POTTER. Rev. ed. Philadelphia: Chilton Book Co., 1973. xix, 330 p. Illus. (part color), tables, charts. Glossary, pp. 306-9. Appendix, pp. 310-24. Bibliography, p. 325. Index, pp. 327-30. (Original edition: Chamberlin, no. 1785).

> This standard reference text for students and craftspeople treats most of the technical facets of ceramic work, from the composition of clays to the study of glaze calculation. Rhodes's emphasis is on the technical nature of the materials, particularly in the areas of glazing and firing, rather than on the working of the clay by hand or on the potter's wheel. A selection of glaze formulas is contained in the appendix.

______. KILNS: DESIGN, CONSTRUCTION, AND OPERATION. Philadelphia: Chilton Book Co., 1968. 241 p. 203 illus. Bibliography, pp. 233-34. Index, pp. 235-40.

> Writing perhaps the most complete book available on kilns, Rhodes discusses their development through the centuries and explains their principles of construction and operation. He devotes the final section of his text to informing us of the variety of kiln designs.

______. STONEWARE AND PORCELAIN: THE ART OF HIGH-FIRED POTTERY. Philadelphia: Chilton Book Co., 1959. 217 p. 82 halftones, figs., tables. Index, pp. 213-17.

> Primarily interested, in this volume, in conveying technical information on the clays, glazes, firing, and kilns of high-fired stoneware and porcelain, Rhodes also includes a chapter on the historical tradition of these wares. He describes, too, the rhythms of working in the studio and supports his text with a portfolio of illustrations of historical and contemporary objects.

Riegger, Hal. PRIMITIVE POTTERY. New York: Van Nostrand Reinhold Co., 1972. 120 p. Illus. (part color and figs.). Bibliography, p. 119. Index, p. 120.

> Well illustrated and including a good deal of technical information on the forming and firing methods of nontechnical societies

throughout the world, Riegger's text seeks to reawaken sensitivity and awareness so that the reader can work with clay without the elaborate tools of modern technology.

_____. RAKU: ART AND TECHNIQUE. New York: Van Nostrand Reinhold Co., 1970. 136 p. Illus. (part color and figs.). Bibliography, p. 134. Index, pp. 135-36.

Concerned with the philosophy and techniques of raku, Riegger's text deals clearly with its material on clay bodies, glazes, and the building and firing of raku kilns. Riegger limits his details to encourage his reader to more personal experimentation.

Rosenthal, Ernst. POTTERY AND CERAMICS: FROM COMMON BRICK TO FINE CHINA.

See: Chapter 2: "General Histories, Dictionaries, and Encyclopedias" (p. 12).

*Rothenberg, Polly. THE COMPLETE BOOK OF CERAMIC ART. New York: Crown Publishers, 1972. 276 p. Illus. (incl. 18 color). Bibliography, p. 263. Glossary, pp. 264-68. Supply sources, pp. 269-72. Index, pp. 273-76.

The focus of this fully illustrated text is on forming methods and decoration although raku, jewelry, ceramic sculpture, and architecture are also covered. While this volume does encompass a wide variety of material, its claim to being "complete" is perhaps not entirely accurate because it is impossible to cover all the areas the book does in great detail. Both glazing and kiln firing, for example, are treated briefly.

Rottger, Ernst. CREATIVE CLAY DESIGN. Creative Play Series, no. 3. New York: Reinhold Publishing Corp., 1963. 95 p. 254 illus. (4 color).

Principally composing his book of illustrations, which include 250 examples of student work, Rottger emphasizes the importance of shape in creating with clay. He includes photographs of clay animals, masks, and bowls as well as examples of pure design, identifying each as the work of an adult or high school student, male or female. A few pages at the end of the book cover essential information on materials, tools, and techniques.

Roy, Vincent A. CERAMICS: AN ILLUSTRATED GUIDE TO CREATING AND ENJOYING POTTERY. New York: McGraw-Hill Book Co., 1959. 278 p. Illus. (incl. 8 color). Bibliography, pp. 270-72. Index, pp. 273-78.

Designed primarily as a beginner's textbook for junior and senior high school, this work touches on a wide range of subjects, including history, design, and industrial production. Roy also provides several pages of literary references to pottery and lists the official boy scout merit badge requirements. Unfortunately, the book is badly dated.

St. Gaudens, Paul, and Jackson, Arthur R. HOW TO MEND CHINA AND BRIC-A-BRAC: AS A HOBBY . . . AS A BUSINESS. Boston: Charles T. Branford Co., 1953. 123 p. Figs.

This small book instructs in the repair of ceramic objects and other materials by a variety of methods, from simple cementing to more complex riveting and restoration. However, the book suffers from its lack of photographic illustrations and presents several repair methods with too little detail.

Sanders, Herbert H. GLAZES FOR SPECIAL EFFECTS. New York: Watson-Guptill Publications, 1974. 151 p. Illus. (part color), tables. Glossary, pp. 143-44. Bibliography, p. 145. Index, pp. 149-51.

With numerous illustrations of special glazes and finished pieces, Sanders describes the techniques used in producing several unusual glaze effects, including crystalline, copper reduction, and ash glazes; luster and raku wares; and underglaze decoration. Tables, formulas, and detailed technical information on the nature of the materials involved, the compositions of the glazes, firing schedules, and the clay bodies used are also provided.

_____. HOW TO MAKE POTTERY. New York: Watson-Guptill Publications, 1974. 143 p. Illus., tables. Bibliography, p. 140. Index, pp. 142-43.

While it does offer some information on glazing and firing, this fully illustrated text, for the most part, describes the different methods of working with clay, including several hand-building methods, casting, and throwing.

Sanders, Herbert H., in collaboration with Kenkichi Tomimoto. THE WORLD OF JAPANESE CERAMICS. Tokyo and Palo Alto, Calif.: Kodansha International, 1968. 267 p. Illus., diagrams, map. Glossary-index, pp. 253-67.

This informative and well-illustrated book deals quite thoroughly with the working techniques of the Japanese potter of the past and of today. The text contains detailed chapters on tools, forming and decorating processes, and production techniques and offers technical glaze information and formulas for the potter.

Searle, Alfred B. AN ENCYCLOPAEDIA OF THE CERAMIC INDUSTRIES. 3 vols. London: Ernest Benn, 1928. Vol. 1: 391 p. Vol. 2: 462 p. Vol. 3: 384 p. Illus. Bibliography, pp. xi-xxvii. (Chamberlin, no. 1786).

Alphabetically arranged with numerous illustrations and thousands of entries, this encyclopedia gives brief explanations of materials, equipment, processes, periods, and styles of ware. Although basically a manual of technical and commercial information, this volume lists a number of entries that identify types of ware and their history.

Sellers, Thomas. THROWING ON THE POTTER'S WHEEL. A CERAMICS MONTHLY Magazine Handbook. Columbus, Ohio: Professional Publications, 1960. 80 p. Illus. Paper.

Accompanied by illustrations of the throwing process and several contemporary examples of thrown pots, this text covers the basic techniques of throwing and explains specific shapes.

______, ed. CERAMIC PROJECTS: SELECT PROJECTS FOR THE CLASSROOM, HOME AND STUDIO. A CERAMICS MONTHLY Magazine Handbook. Columbus, Ohio: Professional Publications, 1963. 64 p. Illus. Paper.

This selection of ceramic projects was originally published as several articles in CERAMICS MONTHLY magazine. The projects provide a very wide range of difficulty, from that of forming a simple clay whistle to the creation of a complex wall screen with wheel-thrown pieces.

Shafer, Thomas. POTTERY DECORATION. New York: Watson-Guptill Publications; London: Pitman Publishing, 1976. 160 p. 34 color pls. 200 halftones. Bibliography, pp. 156-57. Glossary/Index, pp. 158-60.

Since most pottery texts deal with the entire ceramic process, they seldom treat decoration with any real depth. Shafer's text, however, deals exclusively with decoration and describes in detail an extensive variety of sculptural and color techniques used in decorating ceramic surfaces. There are numerous illustrations of the processes used; both historical and contemporary pieces are used to illustrate the multiplicity of possible techniques.

Shaw, Kenneth. SCIENCE FOR CRAFT POTTERS AND ENAMELERS. New York: Drake Publishers, 1972. 261 p. Illus., tables, charts. Glossary, pp. 232-51. Bibliography, pp. 252-53. Index, pp. 254-61.

Intending to introduce the hobbyist and the craftsperson to a more thorough knowledge of ceramic science, Shaw's text deals with the general materials and firing of ceramics and enamels, covering a number of the most troublesome areas in workshops.

Singer, Felix, and German, W.L. CERAMIC GLAZES. Rev. ed. London: Borax Consolidated, 1971. 112 p. Charts, tables, diagrams. 13 appendixes, pp. 75-110. Index, pp. 111-12. Paper.

This edition of Singer's earlier CERAMIC GLAZES has been revised by Mr. German and contains parts of Singer's LOW SOLUBILITY GLAZES. It also contains technical information on the nature of glaze materials, types of glazes, preparation, application, testing, and defects, discussing lead and leadless glazes at some length.

Soldner, Paul. KILN CONSTRUCTION. New York: American Craftsmen's Council, 1965. 32 p. 29 illus. Paper.

Intended as a stimulus for individual craftspeople to experiment with the building of their own kilns, this pamphlet covers most of the basic information necessary for carrying out such a project. This information is not particularly detailed, however, and while a number of kiln forms are touched upon, Soldner's principal example is the catenary arch kiln.

Southwell, B.C. MAKING AND DECORATING POTTERY TILES. New York: Watson-Guptill Publications, 1972. 128 p. 52 illus. (4 color, some figs.). Bibliography, p. 127. Index, p. 128.

This volume offers technical information on the various techniques of making and decorating ceramic tiles. Included are chapters on hand- and factory-made tiles, glaze decoration, silk-screen printing, and the combining of different techniques.

Taylor, Doris W., and Hart, Anne Button. CERAMIC PAINTING STEP-BY-STEP. Princeton, N.J.: D. Van Nostrand Co., 1966. 144 p. 4 color pls., drwgs. Index, pp. 143-44.

This hobbyist's guide includes ten projects with translucent underglaze painting. Step-by-step instructions explain the painting of pine cones, wisteria, and a pink baby rose.

Trevor, Henry. POTTERY STEP-BY-STEP. New York: Watson-Guptill Publications, 1966. 128 p. 220 halftones. School directory, pp. 103-6. Museum directory, pp. 109-12. Suppliers list, pp. 117-18. Bibliography, pp. 121-22. Glossary, p. 125. Index, p. 127.

This is a fully illustrated basic guide, with chapters on forming, decorating, glazing, and firing. While it includes some technical information, the text is directed toward the beginning student.

Turoff, Muriel Pargh. HOW TO MAKE POTTERY AND OTHER CERAMIC WARE. New York: Crown Publishers, 1949. 145 p. Figs. Bibliography, p. 137. Glossary, pp. 134-35. Index, pp. 139-45.

This rather badly dated introduction touches on the basic areas of ceramic work and contains chapters on working with children and the use of ceramics in occupational therapy.

Tyler, Christopher, and Hirsch, Richard. RAKU. New York: Watson-Guptill Publications, 1975. 176 p. 224 illus. (24 color). Bibliography, pp. 172-73. Glossary and index, pp. 174-76.

In addition to giving technical information on clays, glazes, kilns, and firing, Tyler and Hirsch devote chapters of their interesting text to the nature, history, and perception of raku and approaches to its forms. While not extensively detailed, the text does explain forming methods, basic kiln building, and glaze formulas and is profusely illustrated with the work of approximately thirty contemporary craftspeople.

Vance-Phillips, L. BOOK OF THE CHINA PAINTER. The Art Amateur Series. New York: Montague Marks, 1896. 311 p. 150 figs., 7 color pls. Index and glossary, pp. 305-11.

An extensive number of techniques and the various steps used in china painting are explained here in some detail. Although probably much less of this type of work is done today, the actual techniques appear still applicable.

*Villiard, Paul. A FIRST BOOK OF CERAMICS. New York: Funk & Wagnalls, 1969. 175 p. Illus. Glossary, pp. 165-69. Index, pp. 171-75.

Most of this introductory text deals with a variety of forming methods, such as hand building, throwing on the wheel, and slip casting. The building of an electric kiln is also explained in an illustrated chapter.

Webber, Lawrence E. MAKING AN ELECTRIC KILN. 4th ed. Durham: University of New Hampshire Engineering Experiment Station, 1955. 16 p. 6 illus. Paper.

Some of its material may be somewhat dated because of the marketing of a variety of new insulating materials, but this pamphlet generally provides a clear, illustrated description of the building of one's own small-scale (firing chamber nine by eleven by thirteen inches) electric kiln.

Weiss, Harvey. CERAMICS: FROM CLAY TO KILN. New York: Young Scott Books, 1964. 64 p. Illus.

Weiss's text is a very simple, children's book describing, step-by-step, several basic methods of working with clay. The book covers the full process, from wedging to firing, and a number of sample projects are clearly illustrated with photographs and drawings; it is intended for readers aged ten and older.

Wildenhain, Marguerite. POTTERY: FORM AND EXPRESSION. New York: Reinhold Book Corporation for the American Craftsmen's Council, 1959. 157 p. 97 halftones.

This is more than a book of technical information, for its author expresses her personal beliefs in the development of human creativity for the benefit of all, but particularly for the craftsperson. She treats such topics as the position of the craftsperson in today's world, the nature of clay, and the relation of form to many aspects of the potter's craft.

Winkley, David. POTTERY. Pelham Craft Series. London: Pelham Books, 1974. 166 p. 12 color pls., 74 halftones, figs. Glossary, pp. 11-16. Index, pp. 161-64.

Addressing the beginner and the student with some basic knowledge, Winkley explains both the simplest techniques and the ceramic processes requiring more skill. While the book is not overly detailed, its scope of information, on the nature of clays, drying clay, bisque firing, kiln building, and glaze firings, is greater than that of most introductory texts.

Winterburn, Mollie. THE TECHNIQUE OF HANDBUILT POTTERY. New York: Watson-Guptill Publications, 1969. 176 p. Illus. Index, pp. 173-75.

Well illustrated with both historical and contemporary examples, this book introduces the beginner to the more common handbuilding techniques and includes material on decoration, glazes, and kiln firing.

Zorza, Rosemary. POTTERY; CREATING WITH CLAY. Garden City, N.Y.: Doubleday & Co., 1974. 160 p. 2 color pls., halftones, figs. Appendix, pp. 155-56. Glossary, pp. 157-59. Supplies and bibliography, p. 160. Paper.

Intended as a starting point for the beginner and avoiding too much complex or technical material, this text introduces the whole process of clay work, from setting up the workshop to firing an electric kiln. Tools, hand building, throwing, decoration, glazes, and firing are all treated, although much more space is given to throwing than to the hand-building processes. There is also a brief picture essay on the process of raku firing.

Chapter 12

CERAMIC PERIODICALS

(Historical, Contemporary, Industrial, Hobby)

The periodicals comprising this chapter represent an incomplete and elective list of publications that deal exclusively or frequently with ceramics in the fields of art, crafts, and/or industry. Only English-language periodicals presently being published are included. Many other publications in the fields of art, antiques, and/or archaeology contain material on ceramics and much additional information may be obtained from THE ART INDEX, THE READERS' GUIDE TO PERIODICAL LITERATURE, THE STANDARD PERIODICAL DICTIONARY, and ULRICH'S INTERNATIONAL PERIODICALS DIRECTORY (see chapter 1).

HISTORICAL CERAMICS

AMERICAN JOURNAL OF ARCHAEOLOGY (AJA)
Archaeological Institute of America
260 W. Broadway
New York, New York 10013

1855-- . Quarterly; archaeological ceramics. Indexed: CHEMICAL ABSTRACTS, SOCIAL SCIENCES AND HUMANITIES INDEX.

AMERICAN WEDGWOODIAN.
Wedgwood International Seminar
55 Vandam Street
New York, New York 10013

1966-- . Quarterly; historical Wedgwood.

ANTIQUARIES JOURNAL

Oxford University Press	(or)	Oxford University Press
Press Road, Neasden		200 Madison Avenue
London N.W.10, England		New York, New York 10016

(Journal of the Society of Antiquaries of London)

1921-- . Semiannual; art, historical. Indexed: ART AND ARCHITECTURE

TECHNICAL ABSTRACTS, BRITISH ARCHAEOLOGICAL ABSTRACTS, BRITISH HUMANITIES INDEX.

ANTIQUE DEALER & COLLECTORS' GUIDE
City Magazines
1-3 Wine Office Courtfleet Street
London EC4 3AL, England

1932-- . Monthly; art, historical.

ANTIQUE MONTHLY
Boone, Inc.
P.O. Drawer 2
Tuscaloosa, Alabama 35401

(formerly ANTIQUE QUARTERLY)

1967-- . Monthly; art, historical. Also available: Xerox University Microfilms.

ANTIQUES
Straight Enterprises
551 Fifth Avenue
New York, New York 10017

1922-- . Monthly; art, historical. Indexed: ART INDEX, READERS' GUIDE

APOLLO; THE INTERNATIONAL MAGAZINE OF ART AND ANTIQUES
Apollo Publications
720 Fifth Avenue
New York, New York 10019

1925-- . Monthly; art, historical. Indexed: ART INDEX, BRITISH HUMANITIES INDEX.

ART AND ARCHAEOLOGY TECHNICAL ABSTRACTS
New York University
Institute of Fine Arts, Conservation Center
1 East Seventy-eighth Street
New York, New York 10021

(formerly ICC ABSTRACTS)

1955-- . Semiannual; art, technical material. Indexed: CHEMICAL ABSTRACTS.

ARTIBUS ASIAE
New York University
Institute of Fine Arts
1 East Seventy-eighth Street
New York, New York 10021

1925-- . Quarterly; art and archaeology. Index: ART INDEX.

ARTS OF ASIA
Arts of Asia Publications
Metropole Building
57 Peking Road
Kowloon, Hong Kong

1971-- . Bimonthly; art, historical (occasional contemporary material).

BURLINGTON MAGAZINE
Burlington Magazine Publications
Elm House, Elm Street
London, W.C.1; England

1903-- . Monthly; art, historical. Indexed: ART INDEX, BRITISH HUMANITIES INDEX.

CONNOISSEUR
National Magazine Company
Chestergate House
Vauxhall Bridge Road
London, SW1V 1HF, England

1901-- . Monthly; art, historical. Indexed: ART INDEX, BRITISH HUMANITIES INDEX.

COUNTRY LIFE
I P C Magazines
Tower House, Southampton Street
London, WC2E 9QX, England

1897-- . Weekly; art, historical. Indexed: ART INDEX.

ORIENTAL ART; DEVOTED TO THE STUDY OF ALL FORMS OF ORIENTAL ART
Oriental Art Magazine
12 Ennerdale Road
Richmond, Surrey, England

1955-- . Quarterly; art, historical. Indexed: ART INDEX.

POTTERY COLLECTORS NEWSLETTER
Box 446
Asheville, North Carolina 28802

Monthly; art, historical.

POTTERY SOUTHWEST
Albuquerque Archaeological Society
6207 Mossman Place, N.E.
Albuquerque, New Mexico 87110

1974-- . Quarterly; archaeological ceramics.

SCIENTIFIC PAPERS ON JAPANESE ANTIQUES AND ART CRAFTS
Tokyo National Research Institute of Cultural Properties
Association of Scientific Research of Antiques
13-27 Ueno Park
Taito-ku, Japan

Monthly; historical, technical.

TRANSACTIONS OF THE ENGLISH CERAMIC CIRCLE
W. & J. Mackay
Lordswood, Chatham
Kent, England

(known as THE ENGLISH PORCELAIN CIRCLE before 1933)

Annual; art, historical.

WEDGWOOD SOCIETY OF LONDON. PROCEEDINGS
Wedgwood Society, c/o Mrs. T.B. Jarvis
Romanvilla, Rockbourne
Fordingbridge, Hampshire, England

Biennial; art, historical.

CONTEMPORARY CERAMICS

AFRICAN ARTS
African Studies Center
University of California
Los Angeles, California 90024

1967-- . Quarterly; art, historical, and contemporary. Indexed: ART INDEX.

AMERICAN INDIAN ART
7045 Third Avenue
Scottsdale, Arizona 85251

Quarterly; art, historical and contemporary.

ARTISAN CRAFTS
O. & E. Enterprises
Star Route 4, Box 179F
Reeds Spring, Missouri 65737

1973-- . Quarterly; crafts (all mediums).

CERAMIC REVIEW
Craftsmen Potters Association of Great Britain
17a Newburgh Street
London, W.1, England

1970-- . Bimonthly; crafts (includes some technical information).

CERAMICS MONTHLY
Professional Publications
Box 12248
Columbus, Ohio 43212

1953-- . Monthly; contemporary ceramic design, technical material (occasional historical material). Indexed: ART INDEX, READERS' GUIDE.

CRAFT CONNECTION
Minnesota Crafts Council
1105 Tenth Street, Southwest
Rochester, Minnesota 55901

(journal of the council)

Quarterly; crafts.

CRAFT HORIZONS
American Crafts Council
44 West Fifty-third Street
New York, New York 10019

1941-- . Bimonthly; crafts design. Indexed: ART INDEX, READERS' GUIDE.

CRAFT/MIDWEST
Box 42, 150 Sherman Road
Northbrook, Illinois 60062

(Formerly WORKING CRAFTSMAN)

1971-- . Quarterly; crafts, professional information.

CRAFT ONTARIO
Ontario Craft Foundation
8 York Street, Seventh Floor
Toronto, Ontario, Canada

Crafts; members' information.

CRAFTS
Crafts Advisory Committee
28 Haymarket
London, SW1Y 4SU, England

1973-- . Bimonthly; crafts design.

CRAFTS CANADA
333 King Street, West
Toronto, Ontario, Canada

Bimonthly; crafts design.

THE CRAFTSMAN'S GALLERY
The Guild of American Craftsmen
Box 645
Rockville, Maryland 20851

Quarterly; crafts design.

DESIGN
Curtis International
1100 Waterway Boulevard
Indianapolis, Indiana 46207

(originally KERAMIC STUDIO)

1899-- . Bimonthly; art, crafts. Indexed: ART INDEX, READERS' GUIDE.

NEW ZEALAND POTTER
P.O. Box 12-162
Wellington North, New Zealand

1958-- . Semiannual; design, technical information.

POTTERY IN AUSTRALIA
Potters' Society of Australia
48 Burton Street
Darlinghurst, New South Wales, Australia

1962-- . Biannual; design, technical material.

POTTERY QUARTERLY
Murray Fieldhouse
Northfields Studio
Northfields, Tring, Hertfordshire, England

1954-- . Quarterly (irregular); design, technical information.

STUDIO POTTER
Daniel Clark Foundation
Box 65
Goffstown, New Hampshire 03045

1972-- . Semiannual; historical and contemporary crafts, technical information.

TACTILE
The Canadian Guild of Potters
100 Avenue Road
Toronto, Ontario, Canada

1967-19?? Bimonthly; crafts. (Ceased publication.)

WESTERN POTTER
British Columbia Guild of Potters
c/o Community Arts Council
315 West Cordova Street
Vancouver, British Columbia, Canada

1965-- . Crafts design.

INDUSTRIAL CERAMICS

AMERICAN CERAMIC SOCIETY. BULLETIN
65 Ceramic Drive
Columbus, Ohio 43214

1922-- . Bimonthly; technical material. Indexed: ENGINEERING INDEX, CHEMICAL ABSTRACTS, METAL ABSTRACTS.

AMERICAN CERAMIC SOCIETY. JOURNAL
65 Ceramic Drive
Columbus, Ohio 43214

1918-- . Bimonthly; technical material. Indexed: CHEMICAL ABSTRACTS, ENGINEERING INDEX, METAL ABSTRACTS, APPLIED SCIENCE AND TECHNOLOGY INDEX.

BRICK AND CLAY RECORD
Cahners Publishing Company
5 South Wabash Avenue
Chicago, Illinois 60603

1892-- . Monthly; technical and professional information. Indexed: CHEMICAL ABSTRACTS, ENGINEERING INDEX.

BRITISH CERAMIC SOCIETY. PUBLICATIONS
Shelton House
Stoke Road
Shelton, Stoke-on-Trent ST4 7LQ, England

(formed by merging of BRITISH CERAMIC SOCIETY. JOURNAL and BRITISH CERAMIC SOCIETY. TRANSACTIONS)

1971-- . 8/year; technical information. Indexed: BRITISH TECHNICAL INDEX, CHEMICAL ABSTRACTS, ENGINEERING INDEX, METAL ABSTRACTS.

Ceramic Periodicals

CANADIAN CERAMIC SOCIETY. JOURNAL
Canadian Ceramic Society
2175 Sheppard Avenue East, Suite 110
Willowdale, Ontario, Canada

1928-- . Monthly; technical material. Indexed: CHEMICAL ABSTRACTS.

LA CERAMICA
Casa Editrice Palazzo Vecchio
via Vittorio Emanuele 155
Florence, Italy

1948-- . Bimonthly; art, historical; industrial.

CERAMIC ABSTRACTS
American Ceramic Society
65 Ceramic Drive
Columbus, Ohio 43214

1922-- . Bimonthly; technical and industrial material. Indexed: CHEMICAL ABSTRACTS.

CERAMIC AGE
Ceramic Age Publications
Box 25555
Cleveland, Ohio 44125

(formerly THE CERAMIST)

1921-- . Bimonthly; technical material.

CERAMIC DATA BOOK
Cahners Publishing Company
5 South Wabash Street
Chicago, Illinois 60603

1921-- . Annual; technical information (both crafts and industrial).

CERAMIC FORUM
O. Hommel Company
Box 475
Pittsburgh, Pennsylvania 15230

1925-- . Bimonthly; crafts, industry. Indexed: CHEMICAL ABSTRACTS.

CERAMIC TEACHING PROJECTS & TRADE NEWS
Scott Advertising & Publishing Company
30595 West Eight Mile Road
Livonia, Michigan 48152

(formerly CERAMIC TRADE NEWS AND CATALOG FILE)

1963-- . Quarterly; crafts and industrial information.

CERAMICS JAPAN
Ceramic Society of Japan
2-22-17 Hyakunin-Cho, Shinjuku-Ku
Tokyo 160, Japan

1966-- . Monthly; industrial material.

INTERCERAM (INTERNATIONAL CERAMIC REVIEW)
Verlag Schmid GmbH.
Kaiser-Josef-Str. 217
Postfach 1722
7800 Freiburg, West Germany

1951-- . Quarterly; industrial material.

MONTHLY REPORT OF THE STATE UNIVERSITY OF NEW YORK COLLEGE OF CERAMICS AT ALFRED UNIVERSITY
State University of New York College of Ceramics at Alfred University
Alfred, New York 14802

1936-- . Monthly; technical information (both crafts and industrial).

HOBBY CERAMICS

ART & ACTIVITIES
Publishers Development Corporation
8150 Central Park Avenue
Skokie, Illinois 60076

1937-- . Monthly (September-June); hobby, education (elementary and middle school).

CERAMIC ARTS AND CRAFTS
Scott Advertising and Publishing Company
30595 West Eight Mile Road
Livonia, Michigan 48152

1955-- . Monthly; instructional articles for the hobbyist.

CERAMIC SCOPE
c/o Mel and Joyce Fiske
6363 Wilshire Boulevard
Los Angeles, California 90048

1964-- . Monthly.

Ceramic Periodicals

LITTLE BLACK BOOK
Daisy Publishing Company
429 Borden Avenue North
Seattle, Washington 98109

1973-- . Annual; molded wares.

POPULAR CERAMICS
Popular Ceramics Publications
6011 Santa Monica Boulevard
Los Angeles, California 90038

1949-- . Monthly.

Chapter 13

CERAMIC ORGANIZATIONS AND SOCIETIES

A number of the societies included here are not devoted exclusively to ceramics. Many small and local crafts societies in the United States are listed in the CONTEMPORARY CRAFTS MARKET PLACE. Many world crafts groups are also identified in the WORLD CRAFTS DIRECTORY (see chapter 1).

AMERICAN CERAMIC SOCIETY
65 Ceramic Drive
Columbus, Ohio 43214

Industrial organization.

AMERICAN CRAFTS COUNCIL
44 West Fifty-Third Street
New York, New York 10019

ARCHAEOLOGICAL INSTITUTE OF AMERICA
260 West Broadway
New York, New York 10013

Allied institutions:

American School of Classical Studies in Athens
American Academy in Rome
American Research Center in Egypt
American Schools of Oriental Research
School of American Research
American Research Institute in Turkey
American Institute of Iranian Studies

ARTIST-CRAFTSMEN OF NEW YORK
130 East Twenty-eighth Street
New York, New York 10016

Successor to the New York Society of Craftsmen and the New York Society of Ceramic Arts.

ASSOCIATION OF SCIENTIFIC RESEARCH OF ANTIQUES
Tokyo National Research Institute of Cultural Properties
13-27 Ueno Park
Taito, Japan

BRITISH CERAMIC SOCIETY
Shelton House
Stoke Road
Shelton, Stoke-on-Trent, England

Industrial organization.

CANADIAN CERAMIC SOCIETY
2175 Sheppard Avenue, East
Suite 110
Willowdale, Ontario, Canada

Industrial organization.

CANADIAN GUILD OF POTTERS
100 Avenue Road
Toronto, Ontario, Canada

CERAMIC SOCIETY OF JAPAN
2-22-17 Hyakunin-Cho, Shinjuku-Ku
Tokyo 160, Japan

Industrial organization.

CONTEMPORARY CRAFTS ASSOCIATION
3934 Southwest Corbett Avenue
Portland, Oregon 97201

CRAFTSMEN POTTERS ASSOCIATION OF GREAT BRITAIN
5 Belsize Lane
London NW3 5AD, England

ENGLISH CERAMIC CIRCLE
14 Dorset Square
London, N.W. 7, England

FEDERATION OF BRITISH CRAFT SOCIETIES
6 Queen Square
London, WCIN 3AR, England

Represents some thirty British craft societies (there are no individual members) and provides information on the work of these groups.

FRIENDS OF THE CRAFTS
311 1/2 Occidental Avenue South
Seattle, Washington 98104

GUILD OF AMERICAN CRAFTSMEN
Box 645
Rockville, Maryland 20851

INTERNATIONAL ACADEMY OF CERAMICS
Consultative Statute of UNESCO
Ariana Museum
Geneva, Switzerland

Contemporary ceramic art.

MINNESOTA CRAFTS COUNCIL
1105 Tenth Street, S.W.
Rochester, Minnesota 55901

NATIONAL COUNCIL ON EDUCATION FOR THE CERAMIC ARTS (NCECA)
Dick Hay, Treasurer
Indiana State University, Art Department
Terre Haute, Indiana 47809

PACIFIC NORTHWEST ARTS & CRAFTS ASSOCIATION
376 Bellevue Square
Bellevue, Washington 98004

POTTER'S SOCIETY OF AUSTRALIA
39 Mary Street
Longueville, New South Wales, Australia

SO-DEI-SHA
6-Chome, Gojohashi, Higashiyama-ku
Kyoto, Japan

Experimental ceramic group (not strictly a society).

THE WEDGWOOD COLLECTOR'S SOCIETY
555 Madison Avenue
New York, New York 10022

Affiliated with the Wedgwood Company and sells new Wedgwood items in limited editions to its members.

THE WEDGWOOD SOCIETY
55 Vandam Street
New York, New York 10013

Ceramic Organizations & Societies

THE WEDGWOOD SOCIETY (Great Britain)
c/o Mrs. T.B. Jarvis
Romanvilla, Rockbourne
Fordingbridge, Hampshire, England

WORLD CRAFTS COUNCIL
29 West Fifty-third Street
New York, New York 10019

Chapter 14

MUSEUM COLLECTIONS IN THE UNITED STATES

My purpose here has been to give the reader a relatively broad survey so that he/she may be alerted to the existence of some small and local collections as well as of those that are large, and even internationally famous. However, I have listed only those institutions that are known to possess enough ceramic objects to warrant the description of their holdings as a collection. Some of this material has been taken from the AMERICAN ART DIRECTORY.

ALABAMA

BIRMINGHAM MUSEUM OF ART
2000 Eighth Avenue North
Birmingham, Alabama 35203

English and continental ceramics; Wedgwood collection; pre-Columbian pottery (middle America and Peru).

ARIZONA

HEARD MUSEUM OF PRIMITIVE ART AND ANTHROPOLOGY
22 East Monte Vista Road
Phoenix, Arizona 85004

Primitive pottery of the world; post-Spanish to contemporary southwestern American Indian pottery (Hohokam, Pima, Papago, Hopi, and Apache).

MUSEUM OF NORTHERN ARIZONA
North Arizona Society of Science and Art
Box 1389
Fort Valley Road
Flagstaff, Arizona 86001

Prehistoric pottery; contemporary American Indian pottery (Hopi, Navaho, Havasupai, etc.).

UNIVERSITY OF ARIZONA MUSEUM OF ART
Olive and Speedway Streets
Tucson, Arizona 85721

American Indian pottery.

CALIFORNIA

M.H. De YOUNG MEMORIAL MUSEUM (now part of THE FINE ARTS MUSEUMS OF SAN FRANCISCO)
Golden Gate Park
San Francisco, California 94118

Oriental ceramics (Avery Brundage collection); Chinese blue-and-white porcelain (Roy C. Leventritt collection).

HUNTINGTON LIBRARY, ART GALLERY AND BOTANICAL GARDENS
San Marino, California 91108

Chelsea porcelain.

LANG ART GALLERY, SCRIPPS COLLEGE (now one of the CLAREMONT COLLEGES)
Ninth and Columbia
Claremont, California 91711

Annual ceramics invitational exhibition.

LOS ANGELES COUNTY ART MUSEUM
5905 Wilshire Boulevard
Los Angeles, California 90036

Egyptian pottery; Greek and Roman pottery; Chinese ceramics; American ceramics.

MILLS COLLEGE MUSEUM
Oakland, California 94613

Contemporary American pottery (Tony Prieto memorial collection).

J. Paul Getty Museum
17985 Pacific Coast Highway
Malibu, California 90265

Greek and Roman vases; French porcelain.

SOUTHWEST MUSEUM
Highland Park
Los Angeles, California 90042

American Indian art (pre-Columbian and post-Columbian).

STANFORD UNIVERSITY MUSEUM
Stanford, California 94305

Ancient pottery (parts of Cesnola collection of Cypriot antiquities).

TAYLOR HALL ART GALLERY
California State University, Chico
Art Department
First Street and Norman Avenue
Chico, California 95926

Eastern ceramics (Behrick Collection).

COLORADO

DENVER ART MUSEUM
100 West Fourteenth Avenue Parkway
Denver, Colorado 80204

American Indian pottery; pre-Columbian pottery; American ceramics; oriental ceramics.

KOSHARE INDIAN MUSEUM
La Junta, Colorado 81050

American Indian pottery.

CONNECTICUT

BRUCE MUSEUM
Bruce Park
Greenwich, Connecticut 06830

Oriental porcelains.

WADSWORTH ATHENEUM
600 Main Street
Hartford, Connecticut 06103

Italian majolica, French and Meissen porcelains (J.P. Morgan collection); pre-Columbian pottery; Eastern ceramics; Bennington ware and American ceramics (Albert Pitkin collection).

YALE UNIVERSITY ART GALLERY
1111 Chapel Street
New Haven, Connecticut 06510

Pre-Columbian pottery (Olsen collection); Greek and Roman vases (Stoddard collection); oriental ceramics; American ceramics.

DELAWARE

THE HENRY FRANCIS DU PONT WINTERTHUR MUSEUM
Winterthur, Delaware 19735

American ceramics (seventeenth to nineteenth centuries); oriental export porcelains made for American market; English and continental pottery and porcelain.

DISTRICT OF COLUMBIA

DUMBARTON OAKS RESEARCH LIBRARY AND COLLECTION (affiliated with Harvard University)
1703 Thirty-second Street, N.W.
Washington, D.C. 20007

Byzantine ceramics; pre-Columbian pottery.

FREER GALLERY OF ART (administered by Smithsonian Institution)
Twelfth and Jefferson Drive, S.W.
Washington, D.C. 20560

Near Eastern ceramics; Chinese, Korean, and Japanese ceramics.

NATIONAL GALLERY OF ART
Constitution Avenue at Sixth Street, N.W.
Washington, D.C. 20565

Index of American Design (twenty-two thousand illustrations of American crafts).

NATIONAL MUSEUM OF HISTORY AND TECHNOLOGY (SMITHSONIAN INSTITUTION)
Fourteenth Street and Constitution Avenue, N.W.
Washington, D.C. 20560

European pottery and porcelain (eighteenth and nineteenth centuries); American pottery (nineteenth century to contemporary); White House china.

FLORIDA

ART MUSEUM OF THE PALM BEACHES
Norton Gallery and School of Art
1451 South Olive Avenue
West Palm Beach, Florida 33401

Chinese ceramics.

CUMMER GALLERY OF ART
DeEtte Holden Cummer Museum Foundation
829 Riverside Avenue
Jacksonville, Florida 32204

Early Meissen porcelain.

FORT LAUDERDALE MUSEUM OF THE ARTS
426 East Las Olas Boulevard
Fort Lauderdale, Florida 33301

American Indian pottery; pre-Columbian pottery.

JACKSONVILLE ART MUSEUM
4160 Boulevard Center Drive
Jacksonville, Florida 32207

Oriental ceramics; pre-Columbian pottery.

GEORGIA

COLUMBUS MUSEUM OF ARTS AND CRAFTS
1251 Wynnton Road
Columbus, Georgia 31906

Oriental Lowestoft and Canton china.

HAWAII

HONOLULU ACADEMY OF ARTS
900 South Beretania Street
Honolulu, Hawaii 96814

Chinese, Japanese, Korean, and Islamic ceramics.

ILLINOIS

ART ASSOCIATION OF JACKSONVILLE
David Strawn Art Gallery
331 West College Avenue
Jacksonville, Illinois 62650

American Indian pottery.

THE ART INSTITUTE OF CHICAGO
Michigan Avenue at Adams Street
Chicago, Illinois 60603

Chinese and Korean ceramics (Russell Tyson collection); English pottery (Frank W. Gunsaulus Wedgwood collection) and porcelain (Amelia Blanxius collection); Greek pottery; pre-Columbian pottery.

FIELD MUSEUM OF NATURAL HISTORY
Roosevelt Road and Lake Shore Drive
Chicago, Illinois 60605

Egyptian pottery; American Indian pottery.

ORIENTAL INSTITUTE OF THE UNIVERSITY OF CHICAGO
1155 East Fifty-eighth Street
Chicago, Illinois 60637

Near Eastern ceramics.

INDIANA

INDIANAPOLIS MUSEUM OF ART
1200 West Thirty-eighth Street
Indianapolis, Indiana 46208

Pre-Columbian pottery; Chinese ceramics.

SOUTH BEND ART ASSOCIATION
South Bend Art Center
121 North Lafayette Boulevard
South Bend, Indiana 46601

Contemporary ceramics.

IOWA

CENTRAL IOWA ART ASSOCIATION
Fisher Community Center
Marshalltown, Iowa 50158

Ceramic study collection--contemporary ceramics, traditional Japanese ceramics.

KANSAS

WICHITA ART ASSOCIATION
9112 East Central
Wichita, Kansas 67206

Contemporary ceramics (from former Wichita Annual Ceramics exhibitions).

WICHITA ART MUSEUM
619 Stackman Drive
Wichita, Kansas 67203

European and oriental porcelains and faience (Florence N. Evans collection).

KENTUCKY

J.B. SPEED ART MUSEUM
2035 South Third Street
Louisville, Kentucky 40208

Eighteenth-century porcelain.

LOUISIANA

NEW ORLEANS MUSEUM OF ART
ISAAC DELGADO MUSEUM
Lelong Avenue, City Park
New Orleans, Louisiana 70179

Pre-Columbian pottery; colonial Latin American pottery; Greek and Roman pottery.

R.W. NORTON ART GALLERY
4747 Creswell Avenue
Shreveport, Louisiana 71106

Wedgwood collection.

MAINE

HAYSTACK MOUNTAIN SCHOOL OF CRAFTS
Deer Isle, Maine 04627

Contemporary American ceramics.

MARYLAND

WALTERS ART GALLERY
600 North Charles Street
Baltimore, Maryland 21201

Sevres porcelain.

MASSACHUSETTS

AMERICAN ANTIQUARIAN SOCIETY
Salisbury Street and Park Avenue
Worcester, Massachusetts 01609

Staffordshire pottery; American ceramics.

MUSEUM OF FINE ARTS, BOSTON
465 Huntington Avenue
Boston, Massachusetts 02115

Greek and Roman pottery; Near Eastern ceramics; Chinese and Korean ceramics; Japanese ceramics (including the Morse collection).

OLD STURBRIDGE VILLAGE
Sturbridge, Massachusetts 01566

American pottery.

PEABODY MUSEUM OF SALEM
161 Essex Street
Salem, Massachusetts 01970

China-trade porcelain; Japanese ceramics (Morse collection).

WILLIAM HAYES FOGG ART MUSEUM
Harvard University
Quincy Street and Broadway
Cambridge, Massachusetts 02138

Greek and Roman vases; Wedgwood.

WILLIAMS COLLEGE MUSEUM OF ART
Main Street
Williamstown, Massachusetts 01267

Mayan pottery.

WORCESTER ART MUSEUM
55 Salisbury Street
Worcester, Massachusetts 01608

Greek vases; Chinese ceramics; pre-Columbian pottery; European ceramics.

MICHIGAN

CRANBROOK ACADEMY OF ART MUSEUM
500 Lone Pine Road
Bloomfield Hills, Michigan 48103

Pre-Columbian pottery; Chinese ceramics; European ceramics; contemporary ceramics.

GREENFIELD VILLAGE AND HENRY FORD MUSEUM
Dearborn, Michigan 48121

English ceramics; American ceramics (seventeenth to twentieth centuries).

MINNESOTA

AMERICAN SWEDISH INSTITUTE
2600 Park Avenue
Minneapolis, Minnesota 55407

Ceramics and china (Scandinavian-related). Porcelain tile fireplaces.

MINNEAPOLIS INSTITUTE OF ARTS
Minnesota Society of Fine Arts
201 East Twenty-fourth Street
Minneapolis, Minnesota 55404

Chinese ceramics; pre-Columbian pottery.

MINNESOTA MUSEUM OF ART
30 East Tenth Street
St. Paul, Minnesota 55101

Contemporary American ceramics; Asian ceramics.

ROURKE ART GALLERY MUSEUM AND ROURKE GALLERY
523 South Fourth Street
Moorhead, Minnesota 56560

American Indian pottery; pre-Columbian Mexican pottery; Islamic pottery.

MISSOURI

ST. LOUIS ART MUSEUM
Forest Park

St. Louis, Missouri 63110

Chinese and Japanese ceramics; Mimbres (American Indian) pottery.

WASHINGTON UNIVERSITY GALLERY OF ART
Steinberg Hall
St. Louis, Missouri 63130

Greek vases.

WILLIAM ROCKHILL NELSON GALLERY OF ART AND MARY ATKINS MUSEUM OF FINE ARTS
4525 Oak Street
Kansas City, Missouri 64111

English pottery (F.P. Burnap collection); Chinese ceramics; American earthenware; Mimbres (American Indian) pottery.

MONTANA

ARCHIE BRAY FOUNDATION
2915 Country Club Avenue
Helena, Montana 59601

Exclusively ceramics: continuously expanding ceramics collection; numerous one-person ceramic exhibitions.

NEW HAMPSHIRE

DARTMOUTH COLLEGE
Hopkins Center Art Galleries
Hanover, New Hampshire 03755

Chinese ceramics.

NEW JERSEY

LENOX CHINA MUSEUM
Trenton, New Jersey 08605

Lenox china.

MONTCLAIR ART MUSEUM
3 South Mountain Avenue
Montclair, New Jersey 07042

Greco-Roman pottery.

NEW JERSEY STATE MUSEUM
Cultural Center
205 West State Street
Trenton, New Jersey 08625

New Jersey ceramics.

PRINCETON UNIVERSITY
The Art Museum
Princeton, New Jersey 08540

World survey of pottery and porcelain covering all periods.

NEW MEXICO

INSTITUTE OF AMERICAN INDIAN ARTS MUSEUM
Cerrillos Road
Santa Fe, New Mexico 87501

Contemporary American Indian pottery.

MAXWELL MUSEUM OF ANTHROPOLOGY
Roma and University Avenue, N.E.
Albuquerque, New Mexico

American Indian pottery.

NEW YORK CITY

AMERICAN MUSEUM OF NATURAL HISTORY
Central Park West at Seventy-ninth Street
New York, New York 10024

Pre-Columbian pottery; American ceramics.

THE BROOKLYN MUSEUM
Eastern Parkway
Brooklyn, New York 11238

Egyptian pottery; pre-Columbian pottery; oriental ceramics; American ceramics; Wedgwood (Emily Winthrop Miles collection).

THE FRICK COLLECTION
1 East Seventieth Street
New York, New York 10021

French porcelain (seventeenth and eighteenth centuries); Chinese ceramics.

HISPANIC SOCIETY OF AMERICA
Broadway at 155th Street
New York, New York 10032

Spanish and Mexican ceramics.

THE METROPOLITAN MUSEUM OF ART
Fifth Avenue at Eighty-second Street
New York, New York 10028

Egyptian pottery; Greek and Roman pottery; Chinese ceramics (Benjamin Altman collection, John D. Rockefeller, Jr., collection); Japanese ceramics; Near Eastern ceramics; European ceramics (R. Thornton Wilson collection); American ceramics.

MUSEUM OF THE AMERICAN INDIAN
Heye Foundation
Broadway at 155th Street
New York, New York 10032

Pre-Columbian pottery; American Indian pottery.

MUSEUM OF CONTEMPORARY CRAFTS OF THE AMERICAN CRAFTS COUNCIL
29 West Fifty-third Street
New York, New York 10019

Twentieth-century American ceramics.

NEW YORK STATE

ALBANY INSTITUTE OF HISTORY AND ART
125 Washington Avenue
Albany, New York 12210

New York ceramics; Chinese export ware; chelsea and bow chinas.

EVERSON MUSEUM OF ART
401 Harrison Street
Syracuse, New York 13202

Seventeenth-, eighteenth-, and nineteenth-century English porcelain; contemporary American ceramics.

JOE AND EMILY LOWE ART CENTER, SYRACUSE UNIVERSITY
309 University Place
Syracuse, New York 13210

Porcelain sculptures by Edward Marshall Boehm.

POTSDAM PUBLIC MUSEUM
Civic Center
Potsdam, New York 13676

English pottery (Burnap collection).

ROBERSON CENTER FOR THE ARTS AND SCIENCES
30 Front Street
Binghamton, New York 13905

Royal Worcester and Meissen porcelains.

Syracuse University Art Collections
Sims Hall
Syracuse, New York 13210

Korean ceramics (John Fox collection); Near-Eastern ceramics.

VASSAR COLLEGE ART GALLERY
Taylor Hall
Poughkeepsie, New York 12601

Far Eastern ceramics (Charles M. Pratt collection); Chinese ceramics (Olga Hasbrouck collection); Greek vases.

NORTH CAROLINA

THE MINT MUSEUM OF ART
501 Hempstead Place
Charlotte, North Carolina 28207

Delhom ceramic collection, pre-Columbian, post-Renaissance, eighteenth-, nineteenth-, and twentieth-century American ceramics.

ST. JOHN'S ART GALLERY
114 Orange Street
Wilmington, North Carolina 28401

Jugtown pottery (American folk pottery).

OHIO

CINCINNATI ART MUSEUM
Eden Park
Cincinnati, Ohio 45202

Chinese ceramics; Near Eastern ceramics; European and American ceramics; Meissen porcelain (Arthur Joseph collection).

CLEVELAND MUSEUM OF ART
11150 East Boulevard
Cleveland, Ohio 44106

European and American ceramics; Chinese and Japanese ceramics.

DAYTON ART INSTITUTE
Forest and Riverview Avenues
Dayton, Ohio 45405

Pre-Columbian pottery; oriental ceramics; Islamic ceramics.

THE DENISON UNIVERSITY ART GALLERY
Burke Hall
Granville, Ohio 43023

Chinese ceramics; American Indian pottery.

OHIO WESLEYAN UNIVERSITY
Department of Fine Arts
Humphreys Art Hall
Delaware, Ohio 43015

Peruvian and Mayan pottery; Chinese and Japanese porcelains; Islamic ceramics.

TAFT MUSEUM
316 Pike Street
Cincinnati, Ohio 45202

Chinese porcelain.

Toledo Museum of Art
2445 Monroe Street
Toledo, Ohio 43697

English and American ceramics of the 18th and 19th centuries (part of the Duckworth collection).

OKLAHOMA

PHILBROOK ART CENTER
2727 South Rockford Road
Tulsa, Oklahoma 74114

American Indian pottery (Clark Field collection).

OREGON

GALLERY 107

Oregon College of Education
Monmouth, Oregon 97361

Contemporary Northwest ceramics.

PORTLAND ART MUSEUM
Portland Art Association
Southwest Park and Madison
Portland, Oregon 97205

Chinese ceramics; contemporary ceramics.

PENNSYLVANIA

BUTEN MUSEUM OF WEDGWOOD
246 North Bowman Avenue
Merion, Pennsylvania 19066

Wedgwood collection.

THE FRICK ART MUSEUM
7227 Reynolds Street
Pittsburgh, Pennsylvania 15208

Eighteenth-century Chinese porcelain.

LANDIS VALLEY FARM MUSEUM
2451 Kissel Hill Road
Lancaster, Pennsylvania 17601

American folk pottery.

MERCER MUSEUM OF THE BUCKS COUNTY HISTORICAL SOCIETY
Pine and Ashland Streets
Doyleston, Pennsylvania 18901

American ceramics (Mercer collection).

MUSEUM OF ART, CARNEGIE INSTITUTE
4400 Forbes Avenue
Pittsburgh, Pennsylvania 15213

Oriental ceramics; Western ceramics.

PHILADELPHIA MUSEUM OF ART
Parkway at Twenty-sixth Street
Philadelphia, Pennsylvania 19130

Tucker porcelain; American ceramics (Gessey collection); oriental ceramics; Near Eastern ceramics.

THE UNIVERSITY MUSEUM
University of Pennsylvania
Thirty-third and Spruce Streets
Philadelphia, Pennsylvania 19174

Ancient pottery from Babylonia, Egypt, Palestine, the Mediterranian, and the Americas.

THE WOODMERE ART GALLERY
9201 Germantown Avenue
Philadelphia, Pennsylvania 19118

European porcelain; oriental porcelain.

RHODE ISLAND

MUSEUM OF ART
RHODE ISLAND SCHOOL OF DESIGN
224 Benefit Street
Providence, Rhode Island 02903

European porcelain (Lucy T. Aldrich collection).

WESTERLY PUBLIC LIBRARY
Broad Street
Westerly, Rhode Island 02891

Lusterware collection.

SOUTH CAROLINA

FLORENCE MUSEUM
Spruce at Graham Streets
Florence, South Carolina 29501

Pueblo ceramics; Chinese ceramics; European ceramics.

TENNESSEE

BROOKS MEMORIAL ART GALLERY
Overton Park
Memphis, Tennessee 38112

Oriental ceramics; porcelain collection.

HOUSTON ANTIQUE MUSEUM
201 High Street
Chattanooga, Tennessee 37403

Porcelains; collection of pitchers--some ceramic.

TEXAS

THE MUSEUM OF FINE ARTS, HOUSTON
1001 Bissonnet Street
Houston, Texas 77005

Egyptian, Greek, and Roman ceramics (Annette Finnigan collection); Southwest American Indian collection (of Miss Ima Hogg); pre-Columbian ceramics (Fabacher-Griffith collection).

VERMONT

BENNINGTON MUSEUM
West Main Street
Bennington, Vermont 05201

Bennington pottery (John Spargo collection, Elizabeth McCullough Johnson collection); Early American pottery (Ruth Loring Whitman collection).

SHELBURNE MUSEUM
United States Route 7
Shelburne, Vermont 05482

American pottery.

VIRGINIA

CHRYSLER MUSEUM AT NORFOLK
Olney Road and Mowbray Arch
Norfolk, Virginia 23510

Ancient pottery of Africa, the Near East, and the Americas; Chinese ceramics.

COLONIAL WILLIAMSBURG
Post Office Drawer C
Williamsburg, Virginia 23185

English pottery and porcelain; American folk art.

VIRGINIA MUSEUM OF FINE ARTS
Boulevard and Grove
Richmond, Virginia 23221

Early Greek vases; pottery and porcelain (Mrs. Arthur Kelly Evans collection); English china (Lady Nancy Astor collection).

WASHINGTON

SEATTLE ART MUSEUM
Volunteer Park
Seattle, Washington 98112

Japanese ceramics (Fuller collection); Chinese ceramics; Near Eastern ceramics; pre-Columbian pottery; eighteenth-century European porcelain (Henry and Martha Isaacson collection).

UNIVERSITY OF WASHINGTON
Henry Art Gallery
Seattle, Washington 98195

Contemporary American ceramics; contemporary Japanese folk pottery; American Indian pottery of the Northwest.

WISCONSIN

CHARLES ALLIS ART LIBRARY
1630 Royall Place
Milwaukee, Wisconsin 53202

(extension of the Milwaukee Public Library)

Chinese, Japanese, Persian, and Korean ceramics (Allis collection).

KENOSHA PUBLIC MUSEUM
Civic Center
Kenosha, Wisconsin 53140

Wisconsin pottery.

MILWAUKEE ART CENTER
750 North Lincoln Memorial Drive
Milwaukee, Wisconsin 53202

American pottery.

THEODORE LYMAN WRIGHT ART CENTER
Beloit College
Beloit, Wisconsin 53511

Korean pottery (Gurley collection).

AUTHOR INDEX

This index is alphabetized letter by letter and numbers refer to page numbers. The names of all contributing authors, editors, and translators are included as well as the names of sponsoring organizations. Only those artists who are contributing authors are indexed here. The names of all ceramic artists discussed or referred to in this bibliography can be found in the subject index.

Author Index

C

O

P

Q

R

TITLE INDEX

This index is alphabetized letter by letter and numbers refer to page numbers. The titles of all books, periodicals, and articles treated in the bibliography are indexed, with the exception of periodicals from which articles are cited. Where two titles are the same, they are differenciated by subtitle or by the author's name in parentheses. Titles of periodicals have been underlined and titles of articles have been placed within quotation marks.

A

Title Index

Title Index

D

E

F

G

H

M

N

O

P

R

S

T

SUBJECT INDEX

This index is alphabetized letter by letter and numbers refer to page numbers. This index is principally organized by country, ceramic type, and historical period. Also indexed are the names of factories, types of wares, museums, and societies as well as all ceramic artists discussed or referred to in the bibliography. Underlined numbers refer to main entries and to entries which seem most valuable in an initial search.

A

Subject Index

F

G

N

W

Y

Z

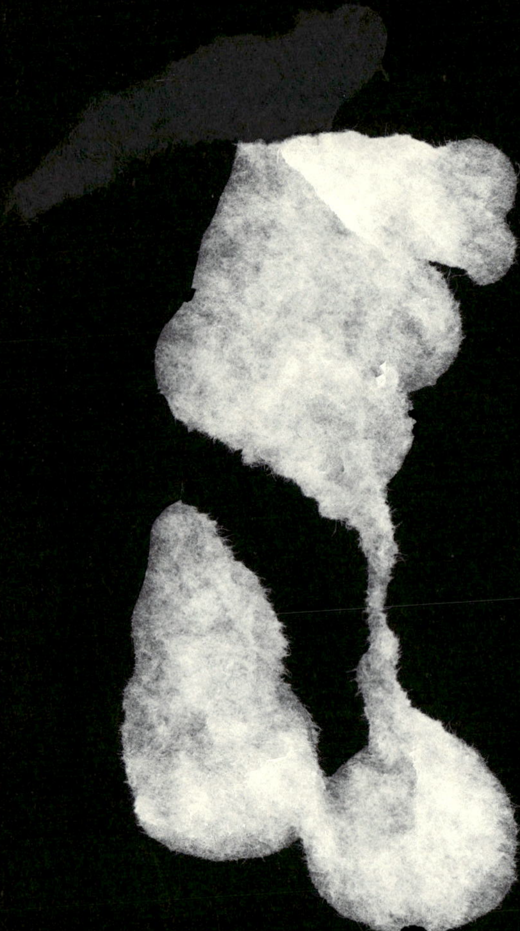